**Bonding and S**
**Molecules and Solids**

# Bonding and Structure
# of Molecules
# and Solids

## D. G. PETTIFOR

*Isaac Wolfson Professor of Metallurgy*
*Department of Materials*
*University of Oxford*

CLARENDON PRESS · OXFORD

Oxford University Press, Walton Street, Oxford OX2 6DP
Oxford New York
Athens Auckland Bangkok Bombay
Calcutta Cape Town Dar es Salaam Delhi
Florence Hong Kong Istanbul Karachi
Kuala Lumpur Madras Madrid Melbourne
Mexico City Nairobi Paris Singapore
Taipei Tokyo Toronto
and associated companies in
Berlin Ibadan

Oxford is a trade mark of Oxford University Press

Published in the United States
by Oxford University Press Inc., New York

First printed 1995
Reprinted (with corrections) 1996

A catalogue record for this book is available from the British Library

Library of Congress Cataloging in Publication Data
Pettifor, D. G. (David G.), 1945–
Bonding and structure of molecules and solids / D.G. Pettifor.
1. Solid state physics. 2. Electronic structure. 3. Molecules.
4. Chemical bonds. 5. Composite materials—Bonding. I. Title.
QC176.P443 1995 541.2'2—dc20 95-32183

ISBN 0 19 851787 4 (Hbk)
ISBN 0 19 851786 6 (Pbk)

Typeset by Integral Typesetting, Gt. Yarmouth, Norfolk
Printed in Great Britain by Bookcraft (Bath) Ltd
Midsomer Norton, Avon

# Preface

'As simple as possible but not simpler'

Albert Einstein

Einstein's caveat is indeed an apt warning for those who wish to explain the structure of molecules and solids within a simple theory or set of rules. The difference in energy between competing structure types is very small, often being a mere hundredth or thousandth of the total cohesive energy. Simple valence bond arguments, for example, can rationalize the well-known 8-N rule for the structures of sp-valent elements by assuming that single covalent bonds are formed between neighbouring atoms, thereby completing the stable octet shell of electrons about each atom. But how do we account for the many exceptions to the 8-N rule such as group IV carbon and lead being most stable in the non-fourfold coordinated graphitic and cubic close-packed structure types respectively? The common assertion that the graphitic structure is stabilized by the formation of $sp^2$ hybrids on the carbon atoms is far too simplistic as the carbon atoms could equally well have formed $sp^3$ hybrids to stabilize the diamond structure instead. In practice, the most stable structure is determined by a delicate balance between opposing terms in the total binding energy.

The *ab initio* prediction of which structure is the most stable by computing and comparing total energies appears at the outset to be a formidable task. The fundamental equation of quantum mechanics, the Schrödinger equation, can be solved exactly for the hydrogen atom. Its solution for the hydrogen molecule and for all other systems is a *many-body* problem. The wave function is now no longer dependent on the coordinates of a single electron but on the coordinates of all the electrons. Unfortunately, the traditional Hartree–Fock approximation to solving the many-body Schrödinger equation was found to be insufficiently accurate for reliable structural predictions to be made for bulk materials, especially metals. A breakthrough occurred, however, in the mid-1960s when Pierre Hohenberg, Walter Kohn, and Liu Sham proved that the total ground state energy of a many-electron system is a functional of the density. This seemingly simple result, by focusing on the electron density rather than the many-body wave function, allowed them to derive an effective *one-electron* Schrödinger equation which could be solved within the so-called local density approximation. Extensive computations during the past two decades have demonstrated the accuracy of local density functional theory in predicting the structural properties of a wide range of ionic, covalent, and metallic systems.

This success of density functional theory allows the whole question of bonding and structure to be formulated within an effective one-electron framework that is so beloved by chemists in their molecular orbital description of molecules and by physicists in their band theory description of solids. In this book I have tried to follow Einstein's dictum by simplifying the one-electron problem to the barest essentials necessary for understanding observed trends in bonding and structure. In particular, the chemically intuitive tight binding approximation is shown to provide a unified treatment of the covalent bond in small molecules and extended solids, whereas the physically intuitive nearly free electron approximation is found to give a natural description of the metallic bond in sp-valent metals. Emphasis is placed on recent theoretical developments that link structural stability to local topology or connectivity of the lattice through the moments of the electronic density of states. This moments approach creates a powerful bridge between the physicists' view of the *global* electronic structure in reciprocal space and the chemists' view of *local* bonding in real space. We will see that it leads to a fundamental understanding of the structural trends within the periodic table for the elements and within the AB structure map for binary compounds, experimental trends that are presented and discussed in the first chapter.

This book is directed at final-year undergraduates and first-year postgraduates in physics, chemistry, and materials science who have already attended introductory courses on quantum mechanics and molecular orbital and/or band theory. I assume, therefore, that the reader is familiar with concepts such as the covalent bond, hybrid orbitals, and electronegativity. However, in order to understand the structural trends that are observed amongst the elements and binary compounds, it is necessary not only to quantify these old concepts but also to introduce new concepts that extend our physical and chemical intuition. Both the quantification of old concepts and the development of new concepts requires the reader to take the plunge and to be swept along by the internal logic and predictive power of the simple models presented in this book. Most of the illustrative examples in the text require no more mathematical ability than the solution of a quadratic equation. By working through these examples readers will gain insight and experience that allows the newly learned concepts to become part of their everyday intuition and vocabulary. This intuition may be further honed by problems at the end of the book. The plunge required for understanding the elegant second-order perturbative treatment of the structure of sp-valent metals is deeper and more bracing than that needed for solving a $2 \times 2$ secular equation, so that Chapter 6 could be omitted on a first reading of the book.

This book is the product of many people's input and ideas. The importance of my past and present research students and postdocs is reflected in the credits to many figure captions. I should, however, like to mention in particular Nguyen Manh Duc and Paul Lim who helped with a number of the illustrations. I should also like to acknowledge the very helpful comments of Sir Alan Cottrell, Volker Heine, and John Jefferson on the first draft of this book which had been magnificently typed by Béatrice May from a nearly illegible original text. And finally my warmest thanks to C. T. Liu who was amongst the first to appreciate the beauty and usefulness of the phenomenological structure maps, Masato Aoki who kept faith with the bond order potentials, Adrian Sutton who shares the dream of modelling materials across all the length scales, and Di Gold who provided love and support during the summer of '94 when this book was written.

*Oxford*                                                                                                                D.G.P.
June 1995


I should like to thank Professor W. B. Holzapfel for bring to my attention the recent recommendations of IUPAC regarding the Pearson notation.

*Oxford*                                                                                                                D.G.P.
July 1996

# Contents

# Note on the choice of units

The energy and length scales that are appropriate at the atomic level are those set by the ionization potential and first Bohr radius of the hydrogen atom. In SI units the energy and radius of the $n$th Bohr stationary orbit are given by

$$E_n = -\frac{1}{n^2}\left(\frac{me^4}{32\pi^2\varepsilon_0^2\hbar^2}\right) \tag{1}$$

and

$$a_n = \left(\frac{4\pi\varepsilon_0\hbar^2}{me^2}\right)n^2 \tag{2}$$

where $m$ is the electronic mass, $e$ is the magnitude of the electronic charge, $\varepsilon_0$ is the permittivity of free space, and $\hbar$ is Planck's constant divided by $2\pi$. Substituting in the values $m = 9.1096 \times 10^{-31}$ kg, $e = 1.6022 \times 10^{-19}$ C, $4\pi\varepsilon_0 c^2 = 10^7$, $c = 2.9979 \times 10^8$ m/s and $\hbar = 1.0546 \times 10^{-34}$ Js, we have

$$E_n = -\frac{2.1799 \times 10^{-18}}{n^2} \text{ J} \tag{3}$$

and

$$a_n = 5.2918 \times 10^{-11} n^2 \text{ m} \tag{4}$$

Therefore, the ground state of the hydrogen atom, which corresponds to $n = 1$, has an energy of $-2.18 \times 10^{-18}$ J and an orbital Bohr radius of $0.529 \times 10^{-10}$ m or $0.529$ Å. The first value defines the Rydberg unit (Ry), the latter one the atomic unit (au).

Thus, in atomic units we have

$$E_n = -n^{-2} \text{ Ry} \tag{5}$$

and

$$a_n = n^2 \text{ au} \tag{6}$$

where 1 Ry $= 2.18 \times 10^{-18}$ J $= 13.6$ eV and 1 au $= 5.29 \times 10^{-11}$ m $= 0.529$ Å. It follows from (1), (2), (5), and (6) that $\hbar^2/(2m) = 1$ and $e^2/(4\pi\varepsilon_0) = 2$ in atomic units.

The total energy of the molecule or solid will usually be given in either Ry/atom or eV/atom. Conversion to other units may be achieved by using 1 mRy/atom $= 1.32$ kJ/mol $= 0.314$ kcal/mol.

# 1
# Experimental trends in bonding and structure

## 1.1. Introduction

The crystal structures of the elements display well-known trends within the periodic table. On a broad scale the close-packed structures that are characteristic of the metallic bond on the left-hand side give way to the more open structures of the covalent bond on the right-hand side. On a finer scale, specific trends are observed such as the hcp → bcc → hcp → fcc sequence across the transition-metal series or the 3-fold coordinated graphitic → 4-fold coordinated diamond → 12-fold coordinated fcc sequence down group IV from C to Pb. The crystal structures of isostoichiometric binary compounds also display well-defined trends when the experimental data base is presented within a two-dimensional structure map. In addition to structural domains reflecting close-packed metallic or open covalent bonding, ionic domains of stability appear, such as the NaCl and CsCl regions amongst the alkali halides.

This chapter presents the experimental structures of the elements and binary AB compounds and highlights the dominant structural trends. Each structure type will be characterized not only by one of the fourteen Bravais lattices, which defines the basic building block for *global* three-dimensional periodicity, but also by the *local* coordination polyhedron about each non-equivalent site, which reflects the bonding type. Further, since our understanding of the nature of the chemical bond in solids must be compatible with a theory of the bonding and structure of molecules, the ground-state structural data base on trimers and elemental sp-valent molecules with up to six atoms will be given.

## 1.2. Structures of the elements

Table 1.1 gives the structures of the elements at zero temperature and pressure. Each structure type is characterized by its common name (when assigned), its Pearson symbol (relating to the Bravais lattice and number of atoms in the cell), and its Jensen symbol (specifying the local coordination polyhedron about each non-equivalent site). We will discuss the Pearson and Jensen symbols later in the following two sections. We should note,

## Table 1.1 The ground state structures of the elements

Hydrogen, nitrogen, oxygen, and the halogens bind as dimers that are then weakly held together on the lattices which are indicated by the Pearson symbol. (Data from P. Villars and J.L.C. Daams (1993))

Legend (box):
- Element
- Common name
- Pearson symbol
- Jensen symbol

| Element | Common name | Pearson symbol | Jensen symbol |
|---|---|---|---|
| H$_2$ | | hP2 | 1 |
| He | | hP2 | 12 |
| Li | (Sm) | hR9 | 12,12' |
| Be | | hP2 | 12' |
| B | | hR315 | |
| C | graphite | hP4 | 3' |
| N$_2$ | | hP2 | 1 |
| O$_2$ | | hR6 | 1 |
| F$_2$ | | mS8 | 1 |
| Ne | | cF4 | 12 |
| Na | (Sm) | hR9 | 12,12' |
| Mg | | hP2 | 12 |
| Al | | cF4 | 12 |
| Si | diamond | cF8 | 4 |
| P | black P | oS8 | 3 |
| S | | mP28 | 2 |
| Cl$_2$ | | oS8 | 1 |
| Ar | | cF4 | 12 |
| K | | cI2 | 14 |
| Ca | | cF4 | 12 |
| Sc | | hP2 | 12' |
| Ti | | hP2 | 12' |
| V | | cI2 | 14 |
| Cr | | cI2 | 14 |
| Mn | | cI58 | 12',13,16 |
| Fe | | cI2 | 14 |
| Co | | hP2 | 12' |
| Ni | | cF4 | 12 |
| Cu | | cF4 | 12 |
| Zn | | hP2 | 12' |
| Ga | | oS8 | 7' |
| Ge | diamond | cF8 | 4 |
| As | | hR6 | 6 |
| Se | | hP3 | 2 |
| Br$_2$ | (Cl$_2$) | oS8 | 1 |
| Kr | | cF4 | 12 |
| Rb | | cI2 | 14 |
| Sr | | cF4 | 12 |
| Y | | hP2 | 12' |
| Zr | | hP2 | 12' |
| Nb | | cI2 | 14 |
| Mo | | cI2 | 14 |
| Tc | | hP2 | 12' |
| Ru | | hP2 | 12' |
| Rh | | cF4 | 12 |
| Pd | | cF4 | 12 |
| Ag | | cF4 | 12 |
| Cd | | hP2 | 12' |
| In | | tI2 | 12 |
| Sn | diamond | cF8 | 4 |
| Sb | (As) | hR6 | 6 |
| Te | (Se) | hP3 | 2 |
| I$_2$ | (Cl$_2$) | oS8 | 1 |
| Xe | | cF4 | 12 |
| Cs | | cI2 | 14 |
| Ba | | cI2 | 14 |
| La | | hP4 | 12,12' |
| Hf | | hP2 | 12' |
| Ta | | cI2 | 14 |
| W | | cI2 | 14 |
| Re | | hP2 | 12' |
| Os | | hP2 | 12' |
| Ir | | cF4 | 12 |
| Pt | | cF4 | 12 |
| Au | | cF4 | 12 |
| Hg | (Pa) | tI2 | 14 |
| Tl | | hP2 | 12' |
| Pb | | cF4 | 12 |
| Bi | (As) | hR6 | 6 |
| Po | | cP1 | 6 |
| At | | - | - |
| Rn | | - | - |
| Fr | | - | - |
| Ra | | - | - |

### Lanthanides

| Element | Common name | Pearson symbol | Jensen symbol |
|---|---|---|---|
| La | | hP4 | 12,12' |
| Ce | | cF4 | 12 |
| Pr | (La) | hP4 | 12,12' |
| Nd | (La) | hP4 | 12,12' |
| Pm | (La) | hP4 | 12,12' |
| Sm | | hR9 | 12,12' |
| (Eu) | | | |
| Gd | | hP2 | 12' |
| Tb | | hP2 | 12' |
| Dy | | hP2 | 12' |
| Ho | | hP2 | 12' |
| Er | | hP2 | 12' |
| Tm | | hP2 | 12' |
| (Yb) | | | |
| Lu | | hP2 | 12' |

### Actinides

| Element | Common name | Pearson symbol | Jensen symbol |
|---|---|---|---|
| Ac | | cF4 | 12 |
| Th | | cF4 | 12 |
| Pa | | tI2 | 14 |
| U | | oS4 | 12' |
| Np | | oP8 | 14,16 |
| Pu | | mP16 | 12,14,16 |
| Am | (La) | hP4 | 12,12' |
| Cm | (La) | hP4 | 12,12' |
| Bk | (La) | hP4 | 12,12' |
| Cf | (La) | hP4 | 12,12' |
| Es | | - | - |
| Fm | | - | - |
| Md | | - | - |
| No | | - | - |
| Lr | | - | - |

### Separate boxes

| Element | Common name | Pearson symbol | Jensen symbol |
|---|---|---|---|
| Eu | | cI2 | 14 |
| Yb | | cF4 | 12 |

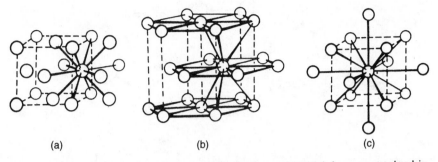

**Fig. 1.1** The three commonest elemental structure types: (a) face-centred cubic, (b) hexagonal close-packed, and (c) body-centred cubic. From Wells (1986).

however, that the periodic table has been arranged slightly differently from usual: the noble gases appear in the first column rather than in the last; Be and Mg have been grouped with Zn, Cd, and Hg rather than with the alkaline earths Ca, Sr, and Ba; and the divalent rare earths Eu and Yb have been assigned under the alkaline earths. We will see that this arrangement of the periodic table leads to a simple scheme for presenting the structural data on isostoichiometric binary compounds within a single two-dimensional structure map.

The most frequently occurring structure types are the close-packed metallic lattices hcp (hexagonal close-packed), fcc (face-centred cubic), and bcc (body-centred cubic) with packing fractions of 0.74, 0.74, and 0.68 respectively, where the packing fraction equals the ratio of the volume occupied by touching hard spheres to the total crystalline volume. Their crystal structures are drawn in Fig. 1.1 where the local bonding arrangement has been emphasized. We see that fcc and hcp have local coordination polyhedra comprising twelve nearest neighbours. On the other hand, bcc has fourteen nearest neighbours as the six second nearest neighbours are only 14% more distant than the eight first nearest neighbours. As is well known the hcp and fcc lattices can be built up by stacking close-packed layers on top of one another in the sequences ABABAB... and ABCABC... respectively, the former being clearly apparent in Fig. 1.1(b). The earlier lanthanides and later actinides take the stacking sequence ABACABAC... whereas samarium takes the sequence [ABABCBCAC]AB..., the structure repeating itself every nine layers. The environment of each successive layer in the La structure type changes from cubic (c) to hexagonal (h) to cubic (c)... so that the stacking sequence can be denoted chchch... whereas in the Sm structure type the sequence reads chhchhchh.... Thus the La structure type can be thought of as being 50% cubic whereas the Sm structure type is only 33%.

The most frequently occurring open structures types are shown in Fig. 1.2.

# 4 Experimental trends in bonding and structure

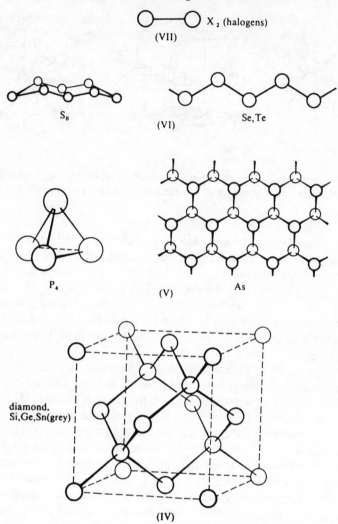

**Fig. 1.2** The commonest open structure types, illustrating the 8-N rule. From Wells (1986).

The halogens in group VII or column 17 of the periodic table crystallize as tightly bound dimers that are held together only weakly by van der Waals forces. They may, therefore, be described as having a local coordination of 1. The chalcogenides in group VI are famous for their numerous polymorphs. Sulphur in its most stable form comprises eight atom puckered rings, whereas selenium and tellurium consist of infinite helical chains. They may be described as having a local coordination of 2, although the interchain

24% longer than the intrachain distance in Se and Te respectively.) The pnictides in group V also form numerous polymorphs. Metastable white phosphorus condenses from the vapour as regular tetrahedral $P_4$ molecules, whereas arsenic, antimony and bismuth take the stable puckered layer structure which is shown in Fig. 1.2. Conventionally the pnictides are assigned a local coordination of 3. (Note that we shall see in Section 1.4 that the interlayer coupling within the arsenic structure is sufficiently strong that As, Sb, and Bi have been assigned a local coordination number of 6 in Table 1.1). Carbon, silicon, germanium, and tin in group IV take the diamond structure with its local coordination of 4. (The most stable form of carbon is, of course, three-fold coordinated graphite.) Ignoring interchain or interlayer coupling, these structure types illustrate the 8-N rule, namely that the number of bonds (or local coordination) equals 8 minus the number of the periodic group. This rule may be rationalized by assuming that single covalent bonds are formed with the neighbours, thereby completing the stable octet shell of electrons about each sp-valent atom.

## 1.3. Lattice types: the Pearson notation

The structure types in Table 1.1 have been characterized by their Pearson symbol which gives the Bravais lattice followed by the number of atoms in the unit cell. There are only 14 different space or Bravais lattices that are compatible with an infinitely repeating three-dimensional crystal. The basic building blocks or unit cells of the Bravais lattices are shown in Fig. 1.3. Depending on their symmetry they may be grouped into seven different crystal systems: triclinic (a), monoclinic (m), orthorhombic (o), tetragonal (t), rhombohedral, hexagonal (h), and cubic (c). The triclinic unit cell is a general parallelepiped with no special relation between the lengths of the sides and angles. The remaining six crystal systems satisfy the symmetry constraints listed in Table 1.2. If there is a lattice point only at the corners of the unit cell, then the Bravais lattice is said to be simple or primitive (P). If there is a lattice point at the centre of the unit cell, then it is said to be body-centred or 'innen' (I). If there is a lattice point at the centre of all three pairs of faces it is said to be face-centred (F), whereas if there is a lattice point at the centre of only one pair of faces it is said to be side face-centred (S). (Note that the symbol S has recently been recommended as standard notation rather than the other commonly used symbol C (Leigh 1990)). The primitive rhombohedral cell in Fig. 1.3 is denoted hR because the primitive rhombohedral Bravais lattice may be thought of as a rhombohedral (R) lattice setting with respect to an underlying hexagonal lattice (h).

The Pearson symbol gives the Bravais lattice with the associated number of atoms in the unit cell. Thus, the simple cubic lattice is designated cP1, whereas the body-centred and face-centred cubic lattices are designated cI2 and cF4 respectively. The primitive rhombohedral Bravais lattice is denoted

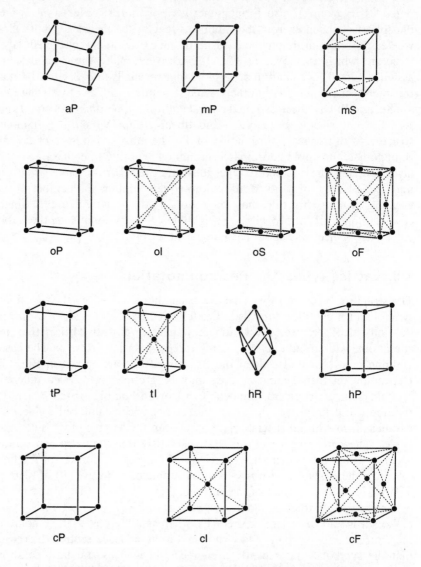

**Fig. 1.3** The fourteen Bravais lattice unit cells and associated lattice symbols.

hR3 because it contains three atoms with respect to the hexagonal unit cell. The diamond lattice, which comprises two interpenetrating fcc lattices, is designated cF8. The hexagonal close-packed lattice is designated hP2, whereas the La and Sm structure types are designated hP4 and hR9 respectively. The complex structures of boron and manganese are designated hR315 and cI58 respectively.

Table 1.2 The fourteen Bravais lattices in three dimensions

| Crystal system | Lattice symbols | Relation between cell edges and angles | Characteristic symmetry |
|---|---|---|---|
| Triclinic | aP | $a \neq b \neq c$ <br> $\alpha \neq \beta \neq \gamma \neq 90°$ | 1-fold (identity or inversion) symmetry only |
| Monoclinic | mP, mS | $a \neq b \neq c$ <br> $\alpha = \gamma = 90° \neq \beta$ | 2-fold axis in one direction only |
| Orthorhombic | oP, oF, oS, oI | $a \neq b \neq c$ <br> $\alpha = \beta = \gamma = 90°$ | 2-fold axes in three mutually perpen-dicular directions |
| Tetragonal | tP, tI | $a = b \neq c$ <br> $\alpha = \beta = \gamma = 90°$ | 4-fold axis in one direction only |
| Rhombohedral | hR | $a = b = c$ <br> $\alpha = \beta = \gamma < 120°, \neq 90°$ | 3-fold axis in one direction only |
| Hexagonal | hP | $a = b \neq c$ <br> $\alpha = \beta = 90°, \gamma = 120°$ | 6-fold axis in one direction only |
| Cubic | cP, cI, cF | $a = b = c$ <br> $\alpha = \beta = \gamma = 90°$ | Four 3-fold axes along body diagonals |

## 1.4. Local coordination polyhedra: the Jensen notation

The Bravais or space lattice does not distinguish between different types of local atomic environments. For example, neighbouring aluminium and silicon both take the same face-centred cubic Bravais lattice, designated cF, even though one is a close-packed twelve-fold coordinated metal, the other an open four-fold coordinated semiconductor. Each structure type in Table 1.1 is, therefore, characterized not only by its Bravais lattice but also by the local coordination polyhedra about each non-equivalent site. These are shown in Fig. 1.4, following the notation suggested by Jensen (1989). He labelled twenty-seven different polyhedra, and Villars et al. (1989) extended the notation to all the other polyhedra which are considered in this chapter. We see, therefore, that the twelve-fold coordinated fcc and hcp lattices are given the labels 12 and 12' respectively. The cuboctahedral arrangement of the former and the twinned cuboctahedral arrangement of the latter are clearly evident from Fig. 1.1(a) and (b) respectively. The twelve-fold co-ordinated icosahedral arrangement of atoms is labelled 12". This is taken by 24 of the 58 atoms in the unit cell of α-Mn. The four-fold coordinated diamond lattice, on the other hand, is designated the Jensen symbol 4, reflecting the tetrahedral coordination which is illustrated in Fig. 1.4.

There is sometimes ambiguity as to which nearest neighbours to assign

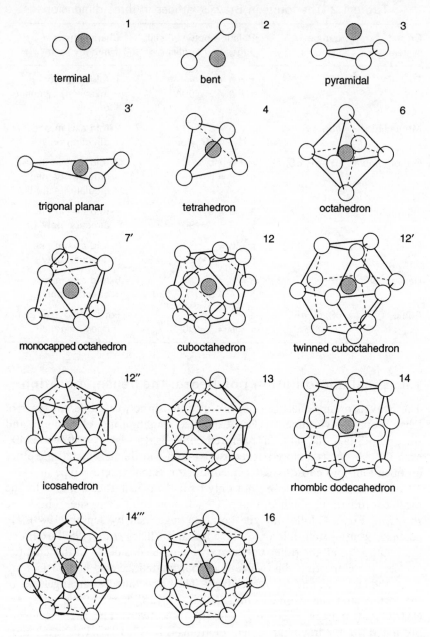

**Fig. 1.4** The local coordination polyhedra and associated Jensen symbols that characterize the elemental ground-state structure types. After Villars and Daams (1993).

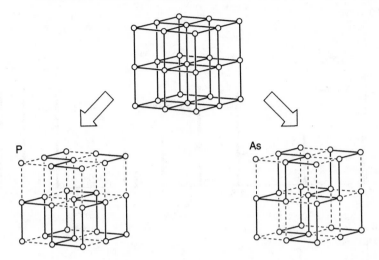

**Fig. 1.5** The breaking of three-bonds about each simple cubic site that leads to the black phosphorus and arsenic structure types. After Burdett and Lee (1985).

to the local coordination polyhedron. This is illustrated by the two different puckered layer structure types which are displayed by the most stable polymorphs of phosphorus and arsenic respectively. They may both be thought of as resulting from the breaking of three bonds about each atom on a simple cubic lattice as shown in Fig. 1.5. The layers then distort. Black phosphorus has two bond angles of 102° and one of 96.5°, whereas arsenic has three bond angles of 96.7°. We saw earlier that the arsenic structure type is conventionally assigned a local coordination number of 3, consistent with the 8-N rule, even though in antimony and bismuth the interlayer distances are only 17% and 12% larger than the intralayer distance.

The assignation of atoms to the local coordination polyhedra in Table 1.1 was made by Villars and Daams (1993) using the maximum-gap rule. Each neighbouring shell is assigned a weight proportional to the number of atoms it contains. A plot of these weights (represented by vertical bars) versus nearest-neighbour distance usually reveals a clear maximum gap, the atoms to the left of the gap forming the coordination polyhedron. As seen in Fig. 1.6 this maximum gap rule gives black phosphorus the three-fold pyramidal coordination 3, whereas arsenic takes the six-fold octahedral coordination 6, characteristic of the simple cubic lattice.

We should note that since the coordination polyhedra reflect the local topology or *connectivity* of an atom, minor distortions of a given polyhedron are designated by the same Jensen symbol. This is an important simplification when seeking order or trends within the very large structural data base. Whereas small distortions in the positions of the atoms within a given crystal can lower the *symmetry* (for example, from cubic to tetragonal),

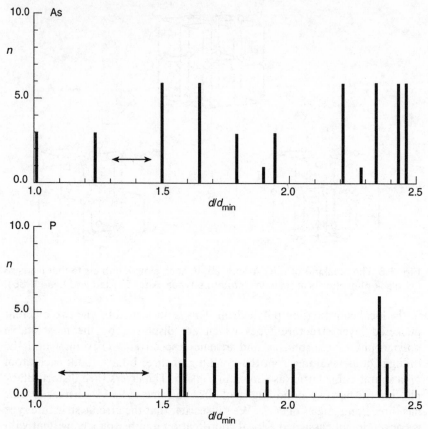

**Fig. 1.6** Black phosphorus and arsenic nearest neighbour histograms, showing the number of atoms in a given neighbouring shell versus the normalized shell distance ($d_{min}$ is the nearest neighbour distance). Note the different location of the 'maximum gap' between phosphorus and arsenic. After Daams *et al.* (1991).

these changes do not usually affect the connectivity and hence the local coordination polyhedron. Thus, face-centred tetragonal indium with an axial ratio $c/a = 1.08$, is labelled in Table 1.1 by the same Jensen symbol 12 as fcc.

## 1.5. Bonding and structural trends within the elements

The variation of the cohesive energy across the periodic table is illustrated by Fig. 1.7. We see the very marked trend across the short periods of the cohesion increasing until the valence shell is half-full at carbon or silicon, thereafter decreasing to almost zero when the valence shell is full for the noble gas solids neon or argon. A similar trend is exhibited across the 4d and 5d transition metal series where the cohesion peaks at niobium and tungsten respectively in the vicinity of a half-full valence d shell, the minima

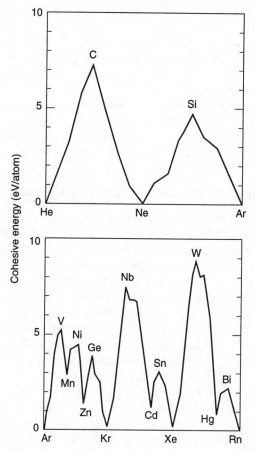

**Fig. 1.7** Cohesive energy across the short periods (upper panel) and long periods (lower panel).

at cadmium and mercury corresponding to the d shell being full and nearly core-like. Surprisingly the largest cohesive energy is not shown by carbon with its strong *saturated* covalent bonds but by the transition metal tungsten. We will see that the bonding in tungsten arises from the strong *unsaturated* covalent bonds between the valence 5d orbitals.

Regarding the structural trends in Table 1.1 we would like theory to be able to explain at least the following:

1. on the broad scale the change from close-packed structures on the left-hand side and middle of the periodic table to the more open structures on the right-hand side;
2. the 8-N rule which gives the number of neighbours expected for $N \geq 4$;
3. the exceptions to the 8-N rule, in particular the fact that the most stable

form of carbon is graphitic (3′), lead is fcc (12), and nitrogen and oxygen are dimeric (1);

4. structural trends within the sp-valent metals, in particular the fact that magnesium is hcp but isovalent calcium and neighbouring aluminium are fcc;
5. the structural trend from hcp → bcc → hcp → fcc across the 4d and 5d transition metal series;
6. the exceptions to the above trend shown by the magnetic 3d transition elements manganese, iron and cobalt;
7. the structural trend from the La structure type [ch] to the Sm structure type [chh] to hcp [h] across the lanthanides.

## 1.6. Bonding and structural trends within AB compounds

The experimental data base on the ground state structures of binary compounds with a given stoichiometry may be displayed within a single two-dimensional structure map. This is achieved by running a one-dimensional string through the two-dimensional periodic table as shown in Fig. 1.8 (where the arrangement of the elements is the same as that in Table 1.1). Pulling the ends of the string apart orders all the elements along a one-dimensional axis, their sequential order defining the relative ordering number $\mathscr{M}$. This simple procedure, which defines a purely phenomenological coordinate, is found to provide an excellent structural separation of all isostoichiometric binary compounds $A_m B_n$ within a single map ($\mathscr{M}_A$, $\mathscr{M}_B$).

The AB structure map where all structure types with four or more representative compounds have been included is given at the end of the book. The Pearson and Jensen symbols for each structure type are given, the latter not being listed when the A or B constituents are associated with more than one local coordination polyhedron. (An exception has been made for TiAs 7/6,6′ since we wish to discuss this structure type later as some arsenic sites have six-fold octahedral coordination 6 whereas others have six-fold trigonal coordination 6′.) Figure 1.9 presents the local coordination polyhedra that have not already been found earlier amongst the elemental ground-state structure types. These include the six-fold coordinated trigonal prism 6′, the seven-fold coordinated pentagonal bipyramid 7, the eight-fold coordinated bicapped trigonal prism 8′′′ and capped octahedron $8^{IV}$, and the nine-fold coordinated tricapped trigonal prism 9.

The bare patches within the structure map indicate that nearly three-quarters of all possible binary AB compounds do not form due to either positive heats of formation or the competing stability of neighbouring phases with different stoichiometry. For example, amongst the transition metals, ScTi, ScV, ScCr, ScFe and ScMn do not form whereas ScCo and ScNi crystallize with the CsCl structure type. We should note that the boundaries between the domains do not have any significance other than they were drawn to separate compounds of different structure type. In regions where

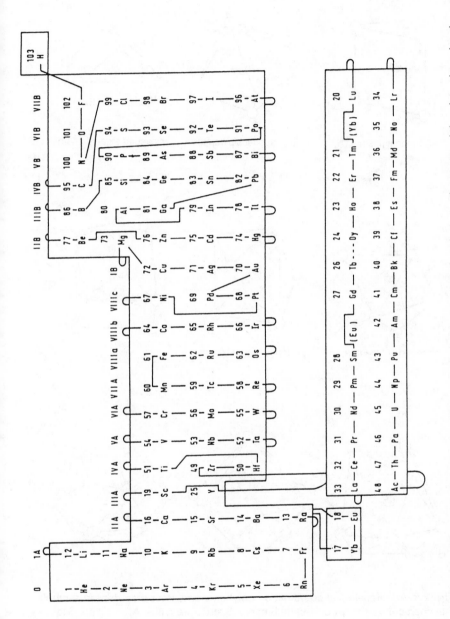

**Fig. 1.8** The string running through this modified periodic table puts all the elements in sequential order, given by the relative ordering number $\mathcal{M}$. From Pettifor (1988).

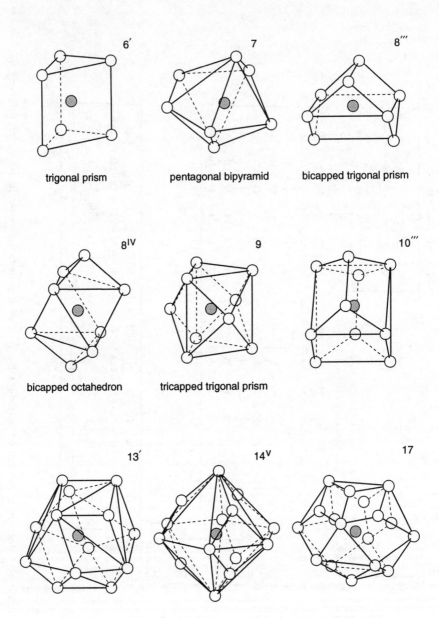

**Fig. 1.9** Additional local coordination polyhedra and associated Jensen symbols that characterize the ground-state binary AB structure types. After Villars, Mathis and Hulliger (1989).

there is a paucity of compounds the boundary is usually chosen as the line separating adjoining groups in the periodic table.

The ten most commonly occurring structure types in order of frequency are NaCl, CsCl, CrB, FeB, NiAs, CuAu, cubic ZnS, MnP, hexagonal ZnS, and FeSi respectively. Structures cF8 (NaCl) and cP2 (CsCl) are ordered with respect to underlying simple cubic and body-centred cubic lattices respectively, as is clear from Figs 1.10(a) and 1.11(a). The Na, Cl sites and Cs, Cl sites are, therefore, six-fold octahedrally coordinated and fourteen-fold rhombic dodecahedrally coordinated, respectively, as indicated by the Jensen symbols 6/6 and 14/14.

Structure tP4 (CuAu) is ordered with respect to an underlying face-centred cubic lattice, so that it takes the Jensen symbol 12/12. The CuAu lattice does show, however, a small tetragonal distortion since the ordering of the copper and gold atoms on alternate (100) layers breaks the cubic symmetry. Zinc blende (cF8(ZnS)) and wurtzite (hP4(ZnS)) are ordered structures with respect to underlying cubic and hexagonal diamond lattices respectively. Since both lattices are four-fold tetrahedrally coordinated, differing only in

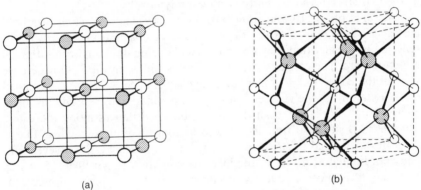

(a)                                         (b)

**Fig. 1.10** The sodium chloride (a) and nickel arsenide (b) structure types. From Wells (1986).

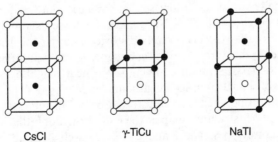

CsCl            γ-TiCu            NaTl

**Fig. 1.11** The cP2(CsCl), tP4(TiCu) and cF16(NaTl) structure types which order with respect to an underlying bcc lattice.

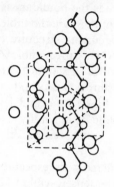

**Fig. 1.12** The iron boride structure type. From Wells (1986).

their dihedral angles, cubic and hexagonal ZnS are labelled by the same Jensen symbols 4/4. We, therefore, expect the energy difference between cubic and hexagonal ZnS structure types to be very small. This is indeed observed as the two structures frequently occur as polymorphs.

Structures oS8(CrB) and oP8(FeB) also have identical Jensen symbols, namely 17/9. Again we expect them to have very similar energies which is reflected in the fact that they occur in neighbouring domains within the AB structure map. The crystal structure of FeB is shown in Fig. 1.12, where we see that the B sites link together in zigzag chains. Each B atom is surrounded by six Fe atoms at the vertices of a trigonal prism, the iron–boron distances being about 20% larger than the boron–boron distances along the chain. Thus, each B site is surrounded by a local coordination polyhedron that is a tricapped trigonal prism, labelled 9 in Fig. 1.9. Each Fe site is seventeen-fold coordinated as given by the local coordination polyhedron 17 in Fig. 1.9. The fifth most frequently occurring structure type, hP4(NiAs), is illustrated in Fig. 1.10(b). We see that each As site is six-fold coordinated by a trigonal prism of Ni atoms. On the other hand, each Ni site is surrounded octahedrally by six As atoms with two near Ni neighbours situated vertically above and below, so that the local coordination polyhedron is a bicapped octahedron. Thus NiAs takes the Jensen symbols $8^{IV}/6'$. The variant of NiAs oP8(MnP) is severely distorted and has the Jensen symbols $10'''/8'''$. The tenth most frequently occurring structure type, cP8(FeSi), is not unrelated to the CsCl structure, having seven near neighbours of opposite type (instead of eight) and six neighbours of like type. The thirteen-fold local coordination polyhedron is illustrated by 13' in Fig. 1.9.

We see in the AB structure map at the end of the book that the two most common structure types, NaCl and CsCl, are well ordered into separate domains. The ionic alkali halides containing NaCl and CsCl are themselves located in the upper left-hand corner of the map. Running across to the right we move from the large cF8(NaCl) 6/6 domain through the sizeable hP4(NiAs) $8^{IV}/6'$ domain to the smaller cF8(ZnS) 4/4 and hP4(ZnS) 4/4 domains. Running down to the bottom we move from the cF8(NaCl) 6/6 domain through the relatively narrow oS8(CrB) 17/9 and oP8(FeB) 17/9 domains to the extensive metallic cP2(CsCl) 14/14 domain. Thus, broadly speaking, the coordination decreases as we move from the top left-hand corner to the right, whereas it increases as we move down to the bottom. This behaviour is consistent with the fact that as we approach the diagonal where $\mathcal{M}_A = \mathcal{M}_B$ we expect to find structures that are similar to those of the elements. Thus, the diagonal runs close to the CsCl, CuAu, and cubic ZnS domains where the elemental bcc, fcc, and diamond lattices are stable.

The Jensen symbols are very important in helping to unravel the relation-ship between the different structure types in neighbouring domains. For example, it is not fortuitous that the NaCl and NiAs domains adjoin each other. Their Jensen symbols 6/6 and $8^{IV}/6'$ tell us immediately that in NaCl the Na and Cl sites are octahedrally coordinated, whereas in NiAs the Ni site is octahedrally coordinated (but with two extra capping atoms), and the As site is trigonally coordinated. It is also not surprising that at the boundary between cF8 (NaCl) 6/6 and hP4(NiAs) $8^{IV}/6'$ we find the two much smaller domains of hP8(TiAs) 7/6, 6' and tI8(NbAs) 6'/6'. Nor is it unexpected to find the two islands of oP8(MnP) $10'''/8'''$ stability in the hP4(NiAs) $8^{IV}/6'$ domain. A distorted NiAs structure type, MnP leads to the bicapping of the trigonal prismatic coordination about the As site, that is $6' \rightarrow 8'''$ (cf Fig. 1.9). Further, we see that the cP8(FeSi) $13'/13'$ domain adjoins a cP2(CsCl) 14/14 domain; they are related structure types as mentioned earlier.

The Jensen notation, however, can say nothing about different possible ordering arrangements of the chemical constituents with respect to a given underlying lattice. For example, CsCl, TiCu, and NaTl all carry the same Jensen symbols 14/14 because they are different ordered structure types with respect to an underlying bcc lattice as shown in Fig. 1.11. The Pearson symbol is then essential for differentiating between the lattices, namely cP2(CsCl), tP4(TiCu), and cF16(NaTl). We see that the AB structure map does an excellent job in separating out these ordered bcc structure types into their respective domains of stability.

We would like theory to be able to explain at least the relative structural trends that are observed amongst the ten most frequently occurring structure types which we have discussed above.

## 1.7. Structural trends within molecules

The concepts required for understanding the bonding and structure of the elements and binary compounds are most easily introduced by considering first the nature of the chemical bond in small molecules. We will use theory

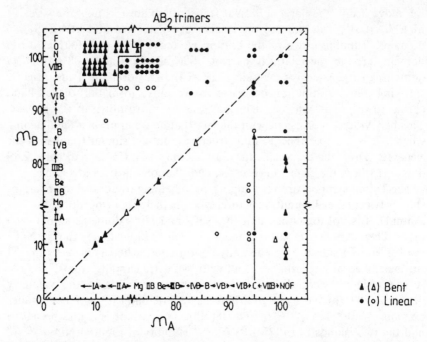

**Fig. 1.13** The Pettifor structure map for sp-valent $AB_2$ triatomic molecules with $N \le 16$ where $N$ is the total number of valence electrons. The full triangles and circles correspond to bent and linear molecules respectively whose shape is well established from experiment or self-consistent quantum mechanical calculations. The open symbols correspond to ambiguous evidence. The data base has been taken from Andreoni *et al.* (1985).

to address the following two questions:

1. Why are some triatomic molecules linear whereas others are bent? For example, carbon dioxide is linear but water is bent with a bond angle of 104.5°. Figure 1.13 shows the structure map for one hundred and eleven sp valent $AB_2$ trimers with $N \le 16$ where $N$ is the total number of valence electrons. We see that good structural separation is obtained, only $Al_2O$ and $Li_2O$ being in the wrong domain.

2. What determines the structural trends amongst the homovalent sp valent molecules with up to six atoms that are displayed in Fig. 1.14? In particular, why amongst the many possible structural variants does $Na_3$ take a bent configuration, $Na_4$ a rhombus, $Na_5$ a two-dimensional close-packed arrangement and $Na_6$ a three-dimensional pentagonal pyramid?

Before addressing these questions we need to remind ourselves of some of the pertinent concepts of quantum mechanics.

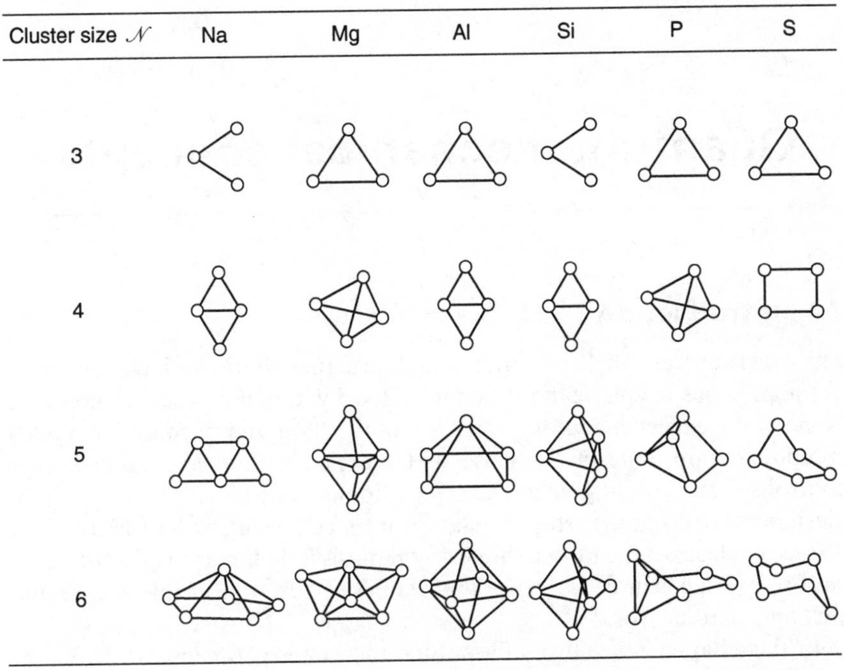

| Cluster size $\mathcal{N}$ | Na | Mg | Al | Si | P | S |
|---|---|---|---|---|---|---|
| 3 | | | | | | |
| 4 | | | | | | |
| 5 | | | | | | |
| 6 | | | | | | |

**Fig. 1.14** The most stable structures of the homovalent 3s-, 3p-valent molecules that are predicted by *ab initio* calculations. Distortions may occur from the idealized structure types drawn. Data compiled by H. S. Lim.

# References

Andreoni, W., Galli, G., and Tosi, M. (1985). *Physical Review Letters* **55**, 1734.

Burdett, J. K. and Lee, S. (1985). *Journal of the American Chemical Society* **107**, 3063.

Daams, J. L. C., Villars, P., and van Vucht, J. H. N. (1991). *Atlas of crystal structure types for intermetallic phases*. ASM International, Materials Park, Ohio.

Jensen, W. B. (1989). *The structures of binary compounds*, Cohesion and structure Vol. 2 (ed. F. R. de Boer and D. G. Pettifor), Chapter 2. North-Holland, Amsterdam.

Leigh, G. J. (ed.) (1990). *Nomenclature of Inorganic Chemistry*, pp. 40, 76. Blackwell, Oxford.

Pettifor, D. G. (1988). *Materials Science and Technology* **4**, 2480.

Villars, P. and Daams, J. L. C. (1993). *Journal of Alloys and Compounds* **197**, 177.

Villars, P., Mathis, K., and Hulliger, F. (1989). *The structures of binary compounds*, Cohesion and structure Vol. 2 (ed. F. R. de Boer and D. G. Pettifor), Chapter 1. North-Holland, Amsterdam.

Wells, A. F. (1986). *Structural inorganic chemistry*. Clarendon Press, Oxford.

# 2
# Quantum mechanical concepts

## 2.1. Introduction

The experimental trends in bonding and structure which we have discussed in the previous chapter cannot be understood within a classical framework. None of the elements and only very few of the thousand or more binary AB compounds are ionic in the sense that the electrostatic Madelung energy controls their bonding. And even for ionic systems, it is a quantum mechanical concept that stops the lattice from collapsing under the resultant attractive electrostatic forces: the strong repulsion that arises as the ion cores start to overlap is direct evidence that Pauli's exclusion principle is alive and well and hard at work!

In this chapter we will briefly outline the key experiments which led to the breakdown of the classical world view and the birth of quantum mechanics. The concept of a wave packet will be used to resolve the problem of wave-particle duality and to underpin Heisenberg's uncertainty principle. A plausibility argument will then be given for the derivation of the Schrödinger equation. It will be solved both for the free-electron gas and free atoms as the former is the starting point for our treatment of the bonding and structure of sp-valent metals, the latter for our treatment of covalent systems. Although this chapter is primarily concerned with introducing the *concepts* behind quantum mechanics, we will end with a short section on the recent *predictive* power of first principles calculations. It is these computationally intensive calculations that have given credence to the simple models which are developed and presented in subsequent chapters.

## 2.2. Wave-particle duality

At the end of the nineteenth century classical physics assumed it had achieved a grand synthesis. The universe was thought of as comprising either matter or radiation as illustrated schematically in Fig. 2.1. The former consisted of point particles which were characterized by their energy $E$ and momentum $\mathbf{p}$ and which behaved subject to Newton's laws of motion. The latter consisted of electromagnetic waves which were characterized by their angular frequency $\omega$ and wave vector $\mathbf{k}$ and which satisfied Maxwell's recently discovered equations. ($\omega = 2\pi v$ and $k = 2\pi/\lambda$ where $v$ and $\lambda$ are the vibrational frequency

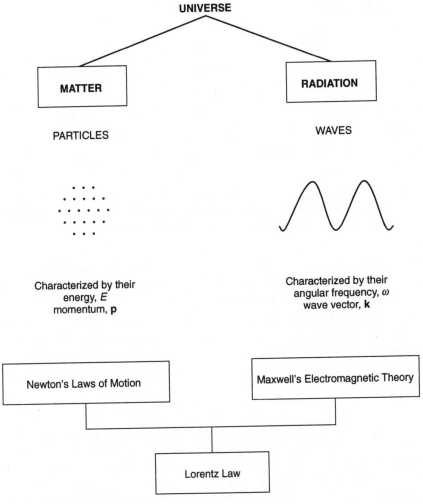

**Fig. 2.1** The grand synthesis of the universe as viewed at the end of the nineteenth century.

and wavelength respectively). The two were coupled together by the Lorentz law $\mathbf{F} = q(\boldsymbol{\varepsilon} + \mathbf{v} \times \mathbf{B})$ where $q$ and $\mathbf{v}$ are the charge and velocity of the point particle, $\boldsymbol{\varepsilon}$ and $\mathbf{B}$ are the electric and magnetic fields, and $\mathbf{F}$ is the resultant force acting on the particle in an electromagnetic field. In this classical description of the universe the point particles then moved under their electromagnetic and gravitational forces according to Newton's laws of motion.

Alas, this grand synthesis soon became unstuck because waves were discovered to display particle-like properties and particles wave-like properties.

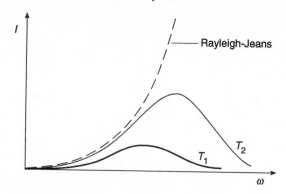

**Fig. 2.2** The intensity of black-body radiation as a function of angular frequency, $\omega$, for two different temperatures, $T_1$ and $T_2$, where $T_2 > T_1$. The dashed curve gives the classical Rayleigh–Jeans law at temperature, $T_2$.

We will consider four key experiments which led to the unravelling of the classical world view, namely black-body radiation, the photoelectric effect, Compton scattering, and electron diffraction.

1. *Black-body radiation*: Figure 2.2 shows the frequency dependence of the intensity of the radiation that is emitted from a black body at two different temperatures, $T_1$ and $T_2$, where $T_2 > T_1$. We see that at high frequencies the emitted intensity, $I$, is much less than that predicted by the Rayleigh–Jeans law, namely

$$I = A\omega^2 k_B T \tag{2.1}$$

where $k_B$ is the Boltzmann constant and $A$ is a given constant. This law follows from the fact that the number density of oscillating modes is proportional to $\omega^2$, and each mode has associated with it an average energy of $k_B T$ by the equipartition of energy, $\frac{1}{2}k_B T$ each for the average kinetic and potential energies of a classical oscillator. The deviation of theory from experiment implied the existence of some barrier to the emission (or absorption) of radiation at higher frequencies. To account for this, Planck assumed in 1900 that electromagnetic energy is not emitted or absorbed continuously but in discrete packets or quanta of energy

$$E = h\nu = \hbar\omega \tag{2.2}$$

where $h$ is Planck's constant and $\hbar = h/2\pi$. A given oscillating mode of angular frequency, $\omega$, then has associated with it a discrete spectrum of allowable energies $E_n = n\hbar\omega$ rather than the continuous spectrum of a classical simple harmonic oscillator. As illustrated schematically in Fig. 2.3 this leads to the high-frequency oscillators not being thermally activated at low temperatures. By Boltzmann statistics, the probability that the $n$th level

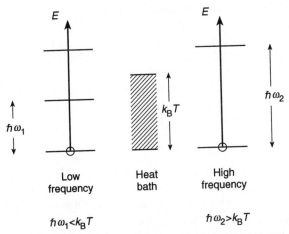

**Fig. 2.3** Schematic illustration of how the quantization of the electromagnetic energy leads to the high-frequency modes being less easily thermally excited than the low frequency.

is excited depends exponentially on its energy, $E_n$, so that the average energy, $\langle E \rangle$, of a quantum oscillator of angular frequency, $\omega$, is given by

$$\langle E \rangle = \left[ \sum_n E_n \, e^{-E_n/k_BT} \right] \Big/ \left[ \sum_n e^{-E_n/k_BT} \right], \tag{2.3}$$

which with $E_n = n\hbar\omega$ simplifies to

$$\langle E \rangle = [(\hbar\omega/k_BT)/(e^{\hbar\omega/k_BT} - 1)]k_BT. \tag{2.4}$$

At low frequencies (or high temperatures) when $\hbar\omega/k_BT \ll 1$, this reduces to the classical limit, $\langle E \rangle = k_BT$, since the prefactor in square brackets tends to unity using the Taylor expansion, $\exp x = 1 + x + \cdots$. At high frequencies (or low temperatures) when $\hbar\omega/k_BT \gg 1$, this reduces to the exponential form, $\langle E \rangle = \hbar\omega \exp(-\hbar\omega/k_BT)$, which accounts for the drop in intensity at high frequencies observed in Fig. 2.2. Fitting Planck's constant to the experimental curves gave $\hbar = 1.055 \times 10^{-34}$ Js. This very small but *non-vanishing* number has important consequences. The comforting image of red hot coals at 1000°K is associated with radiation being emitted at a wavelength of 7000 Å, corresponding to $\hbar\omega/k_BT = 1.8$ eV/0.1 eV = 18. In a classical world, therefore, this red-hot radiation would be more than a million times more intense. At the same time the total energy in the ultraviolet spectrum would diverge catastrophically.

2. *The photoelectric effect*: Light incident on a clean metal surface ejects electrons with energies that depend only on the frequency but not on the intensity of the light. The intensity affects the number of electrons emitted.

This is impossible to understand within a wave description of light since an electromagnetic wave would exert a force on a stationary electron at the metallic surface that is proportional to the electric field and, hence, the electron would be emitted with a velocity dependent on the intensity of the incident light. In 1905 Einstein explained the photoelectric effect by regarding the incident radiation as a beam of particles comprising Planck's quanta with individual energies of $\hbar\omega$. A single light quantum or photon would then kick out an electron from the surface with velocity, $v$, according to

$$\tfrac{1}{2}mv^2 = \hbar\omega - \emptyset \tag{2.5}$$

where $m$ is the electronic mass and $\emptyset$ is the surface work function. Equation (2.5) expresses the conservation of energy; the photon is annihilated during collision, giving up its energy, $\hbar\omega$, to the electron, thus overcoming the work function barrier, $\emptyset$. The electron escapes into the vacuum with kinetic energy, $\tfrac{1}{2}mv^2$. Increasing the intensity of the light increases the number of incident photons. This leads to the emission of a greater number of electrons, but the velocities of these electrons remain unchanged as observed experimentally.

3. *The Compton effect*: A photon is characterized not only by its energy, $E = \hbar\omega$ but also by a momentum, **p**. The latter may be evaluated directly by using the relativistic expression for the energy of a particle in terms of its rest mass, $m$, and momentum, $p$, namely

$$E = (m^2c^4 + p^2c^2)^{1/2}, \tag{2.6}$$

where $c$ is the velocity of light. For a particle at rest, the famous identity, $E = mc^2$, is recovered. For a photon with zero rest mass, $E = pc$. Hence, the momentum of a photon is given by

$$p = \frac{E}{c} = \frac{h\nu}{c} = \frac{h}{\lambda} = \hbar k, \tag{2.7}$$

where we have used $c = \lambda\nu$ and $k = 2\pi/\lambda$, or in vector form

$$\mathbf{p} = \hbar\mathbf{k}. \tag{2.8}$$

In 1923, Compton measured the change in momentum of X-ray photons during scattering by stationary electrons as illustrated schematically in Fig. 2.4. He found that the change in the wavelength of a photon, $\Delta\lambda$, was related to the scattering angle, $\theta$, through the very simple relation

$$\Delta\lambda = 2\lambda_c \sin^2 \frac{\theta}{2}, \tag{2.9}$$

where $\lambda_c$ was a constant. This result could not be understood within classical electromagnetic theory since $\lambda_c$ is then predicted to be a function of the incident wavelength rather than a universal constant. However, eqn (2.9) is

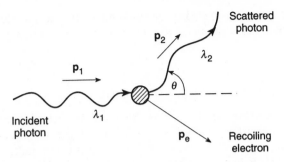

**Fig. 2.4** The scattering of a photon by a stationary electron in the Compton effect.

a direct consequence of applying energy and momentum conservation to the photon-electron scattering processing illustrated in Fig. 2.4. Using the relativistic expression eqn (2.6), the conservation of energy gives

$$cp_1 + mc^2 = cp_2 + (m^2c^4 + p_e^2c^2)^{1/2}, \qquad (2.10)$$

where $p_1$ and $p_2$ are the incident and scattered photon momenta, and $m$ and $p_e$ are the rest mass and scattered momentum of the electron respectively. The conservation of momentum, $\mathbf{p}_1 = \mathbf{p}_2 + \mathbf{p}_e$, gives

$$p_e^2 = p_1^2 + p_2^2 - 2p_1p_2 \cos\theta. \qquad (2.11)$$

Substituting eqn (2.11) into eqn (2.10), and taking $cp_2$ to the left-hand side and squaring yields

$$mc(p_1 - p_2) = 2p_1p_2 \sin^2\frac{\theta}{2}. \qquad (2.12)$$

Dividing by $p_1p_2$ and using $p = h/\lambda$, Compton's result, eqn (2.9), is recovered with $\lambda_c = h/mc$. The term $\lambda_c$ is thus a constant as found experimentally, taking the value of 0.024 Å, the so-called Compton wavelength.

4. *Electron diffraction*: In 1924, de Broglie postulated his principle of wave–particle duality. Just as radiation displays particle-like characteristics, so matter should display wave-like characteristics. It followed, therefore, from eqs (2.2) and (2.7) that a particle with energy, $E$, and momentum, $\mathbf{p}$, has associated with it an angular frequency, $\omega$, and wave vector, $\mathbf{k}$, which are given by

$$\omega = E/\hbar \qquad (2.13)$$

and

$$\mathbf{k} = \mathbf{p}/\hbar. \qquad (2.14)$$

The latter equation implies that a particle of momentum, $p$, has an associated wavelength, $\lambda = h/p$, the de Broglie wavelength. This was verified a year or so later by Davisson and Germer who demonstrated that electrons were diffracted from a nickel crystal as though they had the wavelength postulated by de Broglie.

## 2.3. Heisenberg's uncertainty principle

How do we understand and describe this wave–particle duality? Clearly a plane wave, $A \exp[\mathrm{i}(kx - \omega t)]$, has a well-defined angular frequency, $\omega$ (or energy), and wave vector, $k$ (or momentum). But it is infinite in extent, with its intensity, $|A|^2$, being uniform everywhere in space. In order to create a localized disturbance we must form a wave packet by superposing plane waves of different wave vectors. Mathematically this is written

$$\Psi(x, t) = \frac{1}{\sqrt{2\pi}} \int_{-\infty}^{\infty} \phi(k) \, \mathrm{e}^{\mathrm{i}[kx - \omega(k)t]} \, \mathrm{d}k, \tag{2.15}$$

where the $\phi(k)$ are the amplitudes of the individual plane waves. The angular frequency, $\omega$, is a function of the wave vector, $k$, through the dispersion relation, $\omega = \omega(k)$. At the time, $t = 0$, the wave packet, $\Psi(x, 0)$, is given by

$$\Psi(x, 0) = \frac{1}{\sqrt{2\pi}} \int_{-\infty}^{\infty} \phi(k) \, \mathrm{e}^{\mathrm{i}kx} \, \mathrm{d}k, \tag{2.16}$$

where by Fourier analysis

$$\phi(k) = \frac{1}{\sqrt{2\pi}} \int_{-\infty}^{\infty} \Psi(x, 0) \, \mathrm{e}^{-\mathrm{i}kx} \, \mathrm{d}x. \tag{2.17}$$

Let us consider the simple Gaussian wave packet,

$$\Psi(x, 0) = A \, \mathrm{e}^{\mathrm{i}k_0 x} \, \mathrm{e}^{-(x - x_0)^2/2W^2}, \tag{2.18}$$

which from eqn (2.17) has the associated Gaussian amplitude

$$\phi(k) = AW \, \mathrm{e}^{\mathrm{i}(k_0 - k)x_0} \, \mathrm{e}^{-W^2(k - k_0)^2/2}. \tag{2.19}$$

We see from Fig. 2.5 that the Gaussian wave packet has its intensity, $|\Psi|^2$, centred on $x_0$ with a half width, $W$, whereas $|\phi(k)|^2$ is centred on $k_0$ with a half width, $1/W$. Thus the wave packet, which is centred on $x_0$ with a spread $\Delta x = \pm W$, is a linear superposition of plane waves whose wave vectors are centred on $k_0$ with a spread, $\Delta k = \pm 1/W$. But from eqn (2.8), $p = \hbar k$. Therefore, this wave packet can be thought of as representing a particle that is located approximately within $\Delta x = W$ of $x_0$ with a momentum within $\Delta p = \hbar/W$ of $p_0 = \hbar k_0$. If we try to localize the wave packet by decreasing $W$, we increase the spread in momentum about $p_0$. Similarly, if we try to characterize the particle with a definite momentum by decreasing $1/W$, we increase the uncertainty in position.

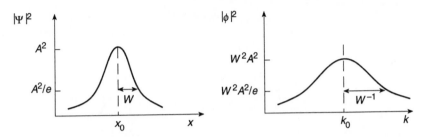

**Fig. 2.5** The relation between a Gaussian wave packet, $\Psi$, and its Fourier transformation, $\phi$. The quantity, $|\Psi|^2$, has a half width of $W$, where $|\phi|^2$ has a half width of $W^{-1}$.

This is a statement of Heisenberg's uncertainty principle, namely

$$\Delta x \times \Delta p \gtrsim \hbar. \tag{2.20}$$

This lies at the heart of the difference between classical and quantum physics. Classically, at any instant in time we can characterize a particle by its exact position, $x$, and exact momentum, $p$, at least in principle. Quantum mechanically, on the other hand, if we know the position $x$, with a high degree of certainty, then there will be a large uncertainty in its momentum, and vice versa.

This uncertainty is an inescapable fact of the world we live in, which is illustrated beautifully by the following *gedanken* experiment suggested by Bohr. Suppose we wish to measure the position of a stationary electron using a microscope as shown in Fig. 2.6. Then from optics the resolution of the microscope will be given by

$$\Delta x \approx \lambda/\sin \theta; \tag{2.21}$$

the resolution is improved by using shorter wavelengths or by increasing the aperture of the lens. But when the stationary electron is illuminated, so that it may be seen, the electron will recoil by the Compton effect. The direction of the photon after scattering is undetermined within the angle subtended by the aperture, so that the photon will have an uncertainty in its momentum in the $x$ direction given by

$$\Delta p_x \approx p \sin \theta \tag{2.22}$$

as seen from Fig. 2.6. Hence, by the conservation of momentum, the magnitude of the recoil momentum of the electron will also be uncertain by

$$\Delta p_x \approx \frac{h}{\lambda} \sin \theta, \tag{2.23}$$

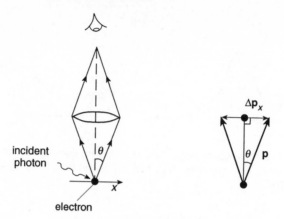

**Fig. 2.6** Schematic illustration of Bohr's *gedanken* experiment. The scattering triangle relates the uncertainty in the momentum of the scattered photon, $\Delta p_x$, to the angle subtended by the aperture, $\theta$, and momentum, $p$, of the photon.

since $p = h/\lambda$ from eqn (2.7). Thus, from eqs (2.21) and (2.23), we have

$$\Delta x \times \Delta p_x \approx h > \frac{h}{2\pi} = \hbar, \tag{2.24}$$

which is consistent with Heisenberg's uncertainty principle.

It is just not possible—in principle—to measure position and momentum with absolute certainty. If we try to determine whether an electron is a wave or a particle, then we find that an experiment which forces the electron to reveal its particle character (for example, one using a very short wavelength microscope) suppresses its wave character as $\Delta p$ and hence $\Delta k$ are large. Alternately, when an experiment focuses on the electron's wavelike behaviour, as in electron diffraction, $\Delta k$ is small, but there is a correspondingly large uncertainty in the position of any given electron within the incident beam.

What does the wave function, $\Psi(\mathbf{r}, t)$, actually represent? Consider the diffraction of an electron beam by two slits as illustrated in Fig. 2.7. Experimentally it is found that the diffraction pattern is not the result of the electrons travelling through slit 1 interfering with those from slit 2. The pattern persists even for the case when the intensity of the incident beam is so low that the particles travel *singly* through the system, being recorded as individual 'flashes' on the screen which build up to give the diffraction pattern observed. This led Born in 1926 to interpret $|\Psi(\mathbf{r}, t)|^2 \, \mathrm{d}\mathbf{r}$ as the probability of finding the electron at some time, $t$, in the volume element $\mathrm{d}\mathbf{r} = \mathrm{d}x \, \mathrm{d}y \, \mathrm{d}z$, which is located at $\mathbf{r} = (x, y, z)$. Since the probability of finding a given electron somewhere in space is unity, the wave function of

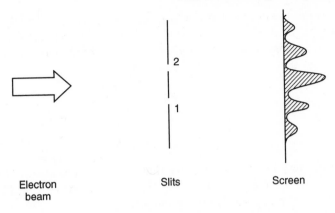

**Fig. 2.7** Diffraction of an electron beam by two slits.

a single electron must satisfy the normalization condition

$$\int_{\text{all space}} |\Psi(\mathbf{r}, t)|^2 \, d\mathbf{r} = 1. \tag{2.25}$$

## 2.4. The Schrödinger equation

In classical mechanics, Newton's laws of motion determine the path or time evolution of a particle of mass, $m$. In quantum mechanics what is the corresponding equation that governs the time evolution of the wave function, $\Psi(\mathbf{r}, t)$? Obviously this equation cannot be obtained from classical physics. However, it can be derived using a plausibility argument that is centred on the principle of wave-particle duality. Consider first the case of a free particle travelling in one dimension on which no forces act, that is, it moves in a region of constant potential, $V$. Then by the conservation of energy

$$E = p^2/2m + V. \tag{2.26}$$

But by de Broglie's principle of wave-particle duality, this free particle has associated with it a wave vector, $k = p/\hbar$, and angular frequency, $\omega = E/\hbar$ (cf eqs (2.13) and (2.14)), which substituting into the above equation gives

$$\hbar\omega = \hbar^2 k^2/2m + V. \tag{2.27}$$

However, this dispersion relation is just that obtained by solving the partial differential equation

$$i\hbar \frac{\partial \Psi(x, t)}{\partial t} = -\frac{\hbar^2}{2m} \frac{\partial^2 \Psi(x, t)}{\partial x^2} + V\Psi(x, t). \tag{2.28}$$

This can be seen directly by substituting into it the plane wave solution,

$$\Psi(x, t) = A \, e^{i(kx - \omega t)}. \tag{2.29}$$

We now make the *ansatz* that the form of this partial differential equation does not change if $V = V(x, t)$ rather than a constant. Generalizing eqn (2.28) to three dimensions, we find the *time-dependent* Schrödinger equation,

$$i\hbar \frac{\partial \Psi}{\partial t} = -\frac{\hbar^2}{2m} \nabla^2 \Psi + V\Psi, \qquad (2.30)$$

where

$$\nabla^2 = \frac{\partial^2}{\partial x^2} + \frac{\partial^2}{\partial y^2} + \frac{\partial^2}{\partial z^2}.$$

We note the very important property that this equation is linear so that if $\Psi_1$ and $\Psi_2$ are solutions, so also is $A_1\Psi_1 + A_2\Psi_2$, generalizing the linear superposition of plane waves we saw earlier.

The time-dependent Schrödinger equation may be solved by the method of separation of variables if the potential does not depend explicitly on time, that is, $V = V(\mathbf{r})$. Writing $\Psi(\mathbf{r}, t) = \psi(\mathbf{r})T(t)$, substituting it into eqn (2.30) and dividing by $\psi(\mathbf{r})T(t)$ we have

$$i\hbar \frac{1}{T} \frac{dT}{dt} = \frac{1}{\psi} \left[ -\frac{\hbar^2}{2m} \nabla^2 \psi + V\psi \right] = \alpha, \qquad (2.31)$$

where $\alpha$ is the usual separation constant. Thus

$$T(t) = e^{-i\alpha t/\hbar} \qquad (2.32)$$

represents an oscillatory function of angular frequency, $\alpha/\hbar$. But by de Broglie's equation (2.13), $\omega = E/\hbar$, so that $\alpha = E$. Hence

$$\Psi_E(\mathbf{r}, t) = \psi_E(\mathbf{r}) e^{-iEt/\hbar}, \qquad (2.33)$$

where from eqn (2.31) the spatially dependent term, $\psi_E(\mathbf{r})$, is an eigenfunction of the *time-independent* Schrödinger equation

$$-\frac{\hbar^2}{2m} \nabla^2 \psi_E(\mathbf{r}) + V\psi_E(\mathbf{r}) = E\psi_E(\mathbf{r}). \qquad (2.34)$$

A system is said to be in a stationary state of energy, $E$, when it is represented by a wave of type eqn (2.33), since then the probability density is independent of time as $|\Psi_E^*(\mathbf{r}, t)\Psi_E(\mathbf{r}, t)| = |\psi_E(\mathbf{r})|^2$.

We should note that the Schrödinger equation is non-relativistic since we derived it from the non-relativistic expression for the energy eqn (2.26). The Dirac equation is the relativistic analogue that is based on the relativistic expression for the energy, namely eqn (2.6). It led directly to the novel concept of electron spin. Since the valence electrons, which control the cohesive and structural properties of materials, usually travel with velocity $v \ll c$, they are adequately described by the Schrödinger equation. For the heavier elements, such as the lanthanides and actinides, relativistic effects can be included perturbatively when necessary. Photons, the quanta of the electromagnetic field, travel at the speed of light and have zero rest mass, so they require the separate treatment of quantum electrodynamics.

## 2.5. The free-electron gas

The sp-valent metals such as sodium, magnesium and aluminium constitute the simplest form of condensed matter. They are archetypal of the textbook metallic bond in which the outer shell of electrons form a gas of free particles that are only very weakly perturbed by the underlying ionic lattice. The classical free-electron gas model of Drude accounted very well for the electrical and thermal conductivities of metals, linking their ratio in the very simple form of the Wiedemann–Franz law. However, we shall now see that a proper quantum mechanical treatment is required in order to explain not only the binding properties of a free-electron gas at zero temperature but also the observed linear temperature dependence of its heat capacity. According to classical mechanics the heat capacity should be temperature-independent, taking the constant value of $\frac{3}{2}k_B$ per free particle.

The Schrödinger equation for a free-electron gas takes the form

$$-\frac{\hbar^2}{2m}\left(\frac{\partial^2}{\partial x^2} + \frac{\partial^2}{\partial y^2} + \frac{\partial^2}{\partial z^2}\right)\psi(\mathbf{r}) = E\psi(\mathbf{r}). \tag{2.35}$$

If the electrons are confined within a box of side $L$, then the normalized eigenfunctions are the plane waves

$$\psi_{\mathbf{k}}(\mathbf{r}) = L^{-3/2}\, e^{i\mathbf{k}\cdot\mathbf{r}}, \tag{2.36}$$

which can be seen by writing $\mathbf{k}\cdot\mathbf{r}$ as $k_x x + k_y y + k_z z$ and substituting into eqn (2.35). The corresponding eigenvalue is given by

$$E = \frac{\hbar^2}{2m}(k_x^2 + k_y^2 + k_z^2) = \frac{\hbar^2}{2m}k^2. \tag{2.37}$$

For a free electron this energy is purely kinetic, so that $E = p^2/2m$. Hence $p = \hbar k = h/\lambda$, as we have already found experimentally for free particles, namely eqs (2.8) and (2.14).

The wavelength, $\lambda$, of the plane wave is constrained by the boundary conditions at the surface of the box. Imposing periodic boundary conditions on the box, which corresponds physically in one dimension to joining both ends of a metallic wire of length, $L$, into a closed ring, we have that $\psi(x + L, y, z) = \psi(x, y, z)$, $\psi(x, y + L, z) = \psi(x, y, z)$, and $\psi(x, y, z + L) = \psi(x, y, z)$. Hence, from eqn (2.36),

$$\mathbf{k} = \frac{2\pi}{L}(n_x, n_y, n_z), \tag{2.38}$$

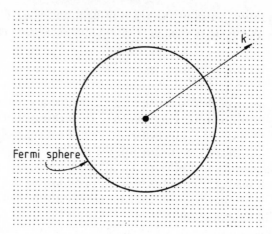

**Fig. 2.8** The fine mesh of allowed **k**-values. At zero temperature only the states **k** within the Fermi sphere are occupied.

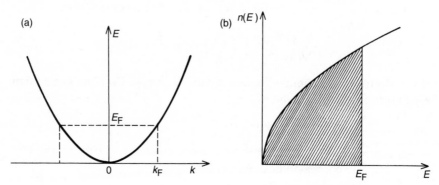

**Fig. 2.9** (a) The free electron dispersion, $E(\mathbf{k})$. (b) The density of states, $n(E)$.

where $n_x$, $n_y$, and $n_z$ are integers. Thus, the allowed values of the wave vector, **k**, are discrete and fall on a fine mesh as illustrated in Fig. 2.8.

Following Pauli's exclusion principle, each state corresponding to a given **k** can contain at most two electrons of opposite spin. Therefore, at the absolute zero of temperature all the states, **k**, will be occupied within a sphere of radius, $k_F$, the so-called Fermi sphere because these correspond to the states of lowest energy, as can be seen from Fig. 2.9(a). The magnitude of the Fermi wave vector, $k_F$, may be related to the total number of valence electrons $N$ by

$$\tfrac{4}{3}\pi k_F^3 2(L/2\pi)^3 = N, \tag{2.39}$$

since it follows from eqn (2.38) that unit volume of **k**-space contains $(L/2\pi)^3$

states capable of holding two electrons each. Thus,

$$k_F = (3\pi^2\rho)^{1/3} = (9\pi/4)^{1/3}/r_s, \qquad (2.40)$$

where $\rho$ is the density of the gas of free electrons, $\rho = N/L^3$, and $r_s$ is the radius of the sphere containing on average one electron, that is

$$\tfrac{4}{3}\pi r_s^3 = L^3/N = \rho^{-1}. \qquad (2.41)$$

The corresponding Fermi energy, $E_F$, equals $\hbar^2 k_F^2/2m$. This takes the values 3.2, 7.1, and 11.6 eV at the electronic densities of sodium, magnesium, and aluminium, which are close to the experimental values for the occupied bandwidths, namely 2.8, 7.6, and 11.8 eV respectively.

The free-electron density of states $n(E)$ may be obtained directly from eqn (2.39) by writing it in the form

$$N(E) = (2m/\hbar^2)^{3/2}(L^3/3\pi^2)E^{3/2}, \qquad (2.42)$$

where $N(E)$ is the total number of states of both spins available with energies less than $E$. Differentiating with respect to energy gives the density of states

$$n(E) = (2m/\hbar^2)^{3/2}(L^3/2\pi^2)E^{1/2}, \qquad (2.43)$$

which is illustrated in Fig. 2.9(b).

We can now see why the experimental electronic heat capacity did not obey the classical result of $\tfrac{3}{2}k_B$ per electron. Following Pauli's exclusion principle, the electrons can be excited into only the unoccupied states above the Fermi energy. Therefore, only those electrons within approximately $k_B T$ of $E_F$ will have enough thermal energy to be excited. Since these constitute about a fraction $k_B T/E_F$ of the total number of electrons we expect the classical heat capacity of $\tfrac{3}{2}k_B N$ to be reduced to the approximate value

$$C_V \approx (T/T_F)\tfrac{3}{2}k_B N, \qquad (2.44)$$

where the Fermi temperature is defined by $E_F = k_B T_F$. (If the calculation is performed exactly, using the square-root dependence of the density of states, then the prefactor changes by the amount, $\pi^2/3$.) For the simple metals magnesium and aluminium, the Fermi temperature is of the order of 100,000 K, so that the electronic heat capacity at room temperature is dramatically reduced compared to the classical prediction for a free-electron gas.

The above model of an sp-valent metal as a gas of free electrons would exhibit no bonding because the only contribution to the energy is the *repulsive* kinetic energy. It takes an average value per electron, which is given by

$$U_{\text{ke}} = \left(\int^{E_F} E n(E)\, \mathrm{d}E\right)\Big/ N = \tfrac{3}{5}E_F. \qquad (2.45)$$

Using eqn (2.40) this may be written explicitly in terms of $r_s$ as

$$U_{ke} = 2.210/r_s^2, \qquad (2.46)$$

where atomic units have been used so that length is measured in units of the first Bohr radius of hydrogen, namely 1 au = 0.529 Å, with energy in units of the ionization potential of hydrogen, namely 1 Ry = 13.6 eV.

We have so far made two implicit assumptions. The first of these is that the gas of electrons is not scattered by the underlying ionic lattice. This can be understood by imagining that the ions are smeared out into a uniform positive background. The second assumption is that the electrons move independently of each other, so that each electron feels the average repulsive electrostatic field from all the other electrons. This field would be completely cancelled by the attractive electrostatic potential from the smeared-out ionic background. Thus, we are treating our sp-valent metal as a metallic jelly or *jellium* within the independent particle approximation.

In practice, however, the motion of the particles is correlated: parallel spin electrons keep apart from each other following Pauli's exclusion principle, and anti-parallel spin electrons keep apart to lower their mutual coulomb repulsion. For interacting electrons the former leads to a lowering in the energy by an amount called the exchange energy, the latter by an amount called the correlation energy. For a homogeneous electron gas, the average exchange-correlation energy per electron is given in atomic units by

$$U_{xc} = -0.916/r_s - (0.115 - 0.0313 \ln r_s), \qquad (2.47)$$

where the first and second terms represent the exchange and correlation energies respectively.

In order to understand this lowering in energy, imagine each electron accompanied by its own mutual exclusion zone or exchange-correlation hole, as illustrated schematically in Fig. 2.10. Each electron now sees an additional attractive potential from the surrounding positive jellium background, which is not screened by electrons. Since the potential at the centre of a sphere of uniform charge varies inversely with the sphere radius, we expect the electron to feel an additional attractive potential proportional to $1/r_s$. This is, indeed, the dominant term in the expression for the exchange-correlation energy, namely eqn (2.47). The latter has been obtained by using the correct form of the exchange-correlation hole, which has a diffuse boundary rather than the sharp boundary assumed in Fig. 2.10.

The resultant binding energy of the free-electron gas,

$$U_{eg} = U_{ke} + U_{xc}, \qquad (2.48)$$

is now attractive, as shown in Fig. 2.11. The maximum cohesion corresponds to a value of 0.16 Ry or 2.2 eV per electron at an effective electronic radius of 4.23 au or 2.2 Å. This is not a bad description of the simple metal sodium, for at equilibrium, it has an experimental cohesive energy of 1.1 eV and an

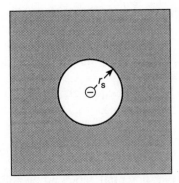

**Fig. 2.10** Schematic illustration of the mutual exclusion zone or exchange-correlation hole about a given electron within a free-electron gas. The hole has a radius, $r_s$, corresponding to exactly one electron being excluded, thereby revealing one positive charge of underlying jellium background. The electron plus its positive hole move together through the gas of other electrons as though they are a neutral entity or quasi-particle.

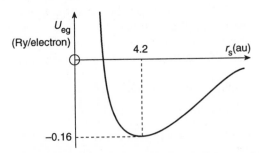

**Fig. 2.11** The binding energy of a free-electron gas as a function of $r_s$.

effective electronic radius of 4.0 au. However, although the other simple metals also have approximately 1 eV per valence electron of binding energy, they show marked variation in their equilibrium atomic volumes and corresponding values of $r_s$. This reflects the discrete size of the ions which has been lost in the jellium model. The important influence of the ion cores on the bonding and structure of the simple metals will be considered in detail in Chapters 5 and 6 respectively.

## 2.6. The free atoms

The free-electron gas model is a good starting point for the sp-valent metals where the loosely bound valence electrons are stripped off from their ion cores as the atoms are brought together to form the solid. However, bonding in the majority of elements and compounds takes place through saturated or unsaturated covalent bonds that form between relatively tightly bound valence orbitals describing, for example, the sp³ hybrids in semiconductors or the d electrons in transition metals. The constituent free atoms then provide the most natural starting point for characterizing the bonding in these tightly bound systems.

The one-electron Schrödinger equation for an isolated free atom takes the form

$$-\frac{\hbar^2}{2m}\nabla^2\psi(\mathbf{r}) + V(r)\psi(\mathbf{r}) = E\psi(\mathbf{r}),$$ (2.49)

where $V(r)$ is an effective potential representing the interaction with the nucleus and the other electrons present (cf eqn (2.61)). Since the potential, $V(r)$, is spherically symmetric the solution takes the separable form

$$\psi_{nlm}(\mathbf{r}) = R_{nl}(r)Y_l^m(\theta, \phi),$$ (2.50)

where $r$, $\theta$, and $\phi$ are spherical polar coordinates and $Y_l^m(\theta, \phi)$ are spherical harmonics. Just as the eigenfunctions for an electron in a cubic three-dimensional box are characterized by the three quantum numbers, $n_x$, $n_y$, and $n_z$, that enter through the boundary conditions (cf eqn (2.38)), so the eigenfunctions of an electron in a spherically symmetric three-dimensional atom are characterized by three-quantum numbers, namely

$n$, the principal quantum number, with $n = 1, 2, 3, \ldots$

$l$, the orbital quantum number, with $l = 0, 1, \ldots, n-1$

$m$, the magnetic quantum number, with $m = 0, \pm 1, \ldots, \pm l$.

The principal quantum number, $n$, determines the energy of the hydrogen atom. The orbital quantum number, $l$, gives the magnitude of the orbital angular momentum, namely $\sqrt{l(l+1)}\hbar$. Electrons with orbital quantum numbers, $l = 0, 1, 2$, and 3 are referred to as s, p, d, and f respectively, after the old terminology of sharp, principal, diffuse, and fine spectroscopic lines. The magnetic quantum number, $m$, specifies the component of the angular momentum in a given direction, taking the $(2l+1)$ values, $m\hbar$, with $|m| \leqslant l$. Because the energy of an electron cannot depend on the direction of its angular momentum in a spherically symmetric potential, these $(2l+1)$ states have the same energy and are said to be degenerate. The magnetic quantum number, $m$, is so called because this degeneracy is lifted in a magnetic field, which fixes a unique direction.

In addition, a relativistic treatment of the electron introduces a fourth quantum number, the spin, $m_s$, with $m_s = \pm\frac{1}{2}$. This is because every electron has associated with it a magnetic moment which it quantized in one of two possible orientations: parallel with or opposite to an applied magnetic field.

The magnitude of the magnetic moment is given by

$$\mu = 2.00\sqrt{s(s+1)}\mu_B, \tag{2.51}$$

where $s = |m_s| = \frac{1}{2}$ and $\mu_B = e\hbar/2m$ (the Bohr magneton). Each eigenstate $nlm$ may, therefore, be doubly occupied by an up ($m_s = +\frac{1}{2}$) and down ($m_s = -\frac{1}{2}$) spin electron. Consequently any given $l$-state of a free atom will be $2(2l+1)$ fold degenerate so that an s-shell can hold 2 electrons, a p-shell 6 electrons, a d-shell 10 electrons, and an f-shell 14 electrons. This electronic shell structure, of course, underpins the periodic table.

The Schrödinger equation can be solved exactly for the case of the hydrogen atom (see, for example, Chapter 12 of Gasiorowicz (1974)). If distances are measured in atomic units, then the first few radial functions take the form

$$R_{1s}(r) = 2e^{-r}, \tag{2.52}$$

$$R_{2s}(r) = \frac{1}{\sqrt{2}}(1 - \tfrac{1}{2}r)\,e^{-r/2}, \tag{2.53}$$

and

$$R_{2p}(r) = \frac{1}{\sqrt{24}}r\,e^{-r/2}. \tag{2.54}$$

They are given by the dashed line curves in Fig. 2.12. A conceptually useful quantity is the probability of finding the electron at some distance, $r$, from the nucleus (in any direction) which is determined by the radial probability density, $P_{nl}(r) = r^2 R_{nl}^2(r)$. We see from the full line curves in Fig. 2.12 that there is a *maximum* probability of locating the electron at the first Bohr radius, $a_1$, for the 1s state and at the second Bohr radius, $a_2$, for the 2p state.

The *average* or expectation value of the radial distance, $r$, is given by

$$\bar{r}_{nl} = [1 + \tfrac{1}{2}(1 - l(l+1)/n^2)]n^2, \tag{2.55}$$

so that $\bar{r}_{1s} = 1.5a_1$, $\bar{r}_{2s} = 1.5a_2$, and $\bar{r}_{2p} = 1.25a_2$. Therefore, the 2s orbital is more extended or diffuse than the corresponding 2p orbital as is evident from Fig. 2.12. This is due to the fact that all solutions of the Schrödinger equation must be orthogonal to one another, that is, if $\psi_{nlm}$ and $\psi_{n'l'm'}$ are any two solutions and $\psi^*$ is the complex conjugate of $\psi$, then

$$\int \psi_{nlm}^* \psi_{n'l'm'} \, d\mathbf{r} = 0. \tag{2.56}$$

If the states have different angular momentum character then the angular integration over the spherical harmonics guarantees orthogonality. But if the states have the same angular momentum character then the orthogonality

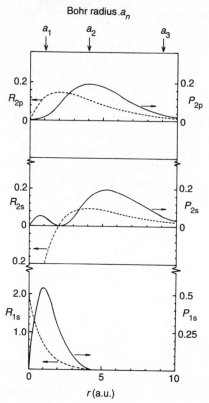

**Fig. 2.12** The radial function, $R_{nl}$ (dashed lines) and the probability density, $P_{nl}$ (solid lines) as a function of $r$ for the 1s, 2s and 2p states of hydrogen.

constraint implies that

$$\int_0^\infty R_{nl}(r)R_{n'l}(r)r^2 \, dr = 0. \tag{2.57}$$

Thus, for the 2s radial wave function to be orthogonal to the 1s radial function, it must change sign, thereby accounting for the node at $r = 2$ au in Fig. 2.12. Similarly, the 3s radial function must be orthogonal to the 2s and, therefore, has two nodes, the 4s has three nodes, etc. Just as the energetically lowest 1s state has no nodes, so the 2p, 3d, and 4f states are nodeless, since they correspond to the states of lowest energy for a given $l$.

The concept of the size of an atom is not well defined within quantum mechanics. An atom has no sharp boundary; the probability of finding an electron decreases exponentially with distance from the atom's centre. Nevertheless, a useful measure of the size of the core is provided by the position of the outer node of the valence electron's radial function, since we have seen that the node arises from the constraint that the valence state

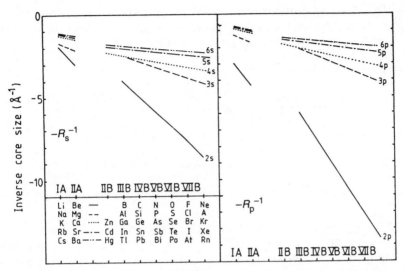

**Fig. 2.13** The negative of the inverse s and p pseudopotential radii of Zunger (1980).

must be orthogonal to the more tightly bound core states. This reflects Pauli's exclusion principle, which states that no two electrons may occupy the same quantum state. A not-unrelated measure of size has been adopted by Zunger (1980), who determined $l$-dependent radii, $R_l$, directly from first principles pseudopotentials (cf Chapter 5).

Figure 2.13 shows his values of $-R_s^{-1}$ and $-R_p^{-1}$ for the sp-bonded elements. We see that the inverse Zunger radii of the *free* atoms vary linearly across the periodic table unlike the internuclear distance in diatomic molecules or the equilibrium atomic volume in bulk systems, which varies parabolically due to the bonding between neighbouring atoms (cf Chapters 3 and 7). As expected, the s and p radii contract across a period as the nuclear charge increases, and they expand down a column as additional full orbital shells are pulled into the core region. One notable exception, however, is given by the crossing of the valence 3s and 4s radii, and also the valence 3p and 4p radii, at group IIIB. This corresponds to Ga having a smaller core than Al due to the fact that the former sits at the end of the 3d transition metal series across which the size of the core has shrunk due to the increase in the nuclear charge. It is not surprising, therefore, that the string through the periodic table in Fig. 1.8 reverses the order of Al and Ga. In addition, we see that the sizes of the second-row elements B, C, N, and O are a lot smaller than those of the other elements in their respective groups, which is again reflected by the string in Fig. 1.8 not linking B, C, N, and O to the isovalent elements in their respective groups.

**Table 2.1** Atomic and ionic radii across the lanthanides. (The former corresponds to a bulk coordination of 12, the latter to an $M^{3+}$ ion with coordination of 6.) (After Douglas *et al.* (1983).)

| Radii (Å) | La | Ce | Pr | Nd | Pm | Sm | Eu | Gd |
|---|---|---|---|---|---|---|---|---|
| Atomic | 1.877 | 1.82 | 1.828 | 1.821 | — | 1.802 | 2.042 | 1.802 |
| Ionic | 1.172 | 1.15 | 1.130 | 1.123 | 1.11 | 1.098 | 1.087 | 1.078 |

| Radii (Å) | Tb | Dy | Ho | Er | Tm | Yb | Lu |
|---|---|---|---|---|---|---|---|
| Atomic | 1.782 | 1.773 | 1.766 | 1.757 | 1.746 | 1.940 | 1.734 |
| Ionic | 1.063 | 1.052 | 1.041 | 1.030 | 1.020 | 1.008 | 1.001 |

Another important example of the influence of core contraction on structural properties is provided by the lanthanides. Table 2.1 gives both their atomic and ionic radii, the former for a bulk coordination of 12, the latter for the triply charged positive ion with a coordination of 6. Firstly, we note that Eu and Yb have much larger atomic volumes than the other lanthanides. This is because they behave as divalent elements in the bulk metal due to the extra stability that arises from a half-full ($f^7$) or completely full ($f^{14}$) shell. We see in Fig. 1.8 that they are separated out and grouped after the divalent alkaline earths. Secondly, we note the marked contraction across the series, lutetium's atomic and ionic radii being 8% and 15% smaller than lanthanum's. This so-called lanthanide contraction causes the subsequent 5d valent element Hf to have a 1% smaller atomic radius than 4d isovalent Zr. This is reflected by the ordering of the string through group IV in Fig. 1.8.

The 4d trivalent element, Y, has an atomic radius that is about midway between that of La and Lu. It is, therefore, not surprising that it is found to be slotted in between Tb and Dy in Fig. 1.8. What is surprising, however, is that the string runs from Lu to La across the series rather than from La to Lu. The latter would have been in the direction of decreasing core size, which is reflected elsewhere in the periodic table as the string moves from left to right. In the last chapter we will see that this apparently anomalous ordering is due to the change in the relative population of the 5d and 6s valence states across the trivalent lanthanides.

The angular character of the orbitals, which is central to directional bonding, is determined by the appropriate spherical harmonic in eqn (2.50). For $m = 0$ the first few spherical harmonics are given by

$$Y_0^0 = \frac{1}{\sqrt{4\pi}}, \tag{2.58}$$

$$Y_1^0 = \sqrt{\frac{3}{4\pi}} \cos \theta, \tag{2.59}$$

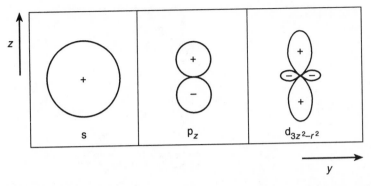

**Fig. 2.14** The angular dependence of the s, $p_z$ and $d_{3z^2-r^2}$ orbitals. The origin of the Cartesian axes passes through the centre of each individual orbital in the directions drawn.

and

$$Y_2^0 = \sqrt{\frac{5}{16\pi}}\,(3\cos^2\theta - 1). \tag{2.60}$$

These are plotted in the $yz$ plane in Fig. 2.14. We see that the s state is spherically symmetric, taking the same positive value in all directions. The p state ($l = 1$, $m = 0$), on the other hand, has a positive lobe pointing out along the positive $z$ direction, but a negative lobe pointing out along the negative $z$ direction. Note that these two lobes are spherical, since the plot, $r = |\cos\theta|$, corresponds to $x^2 + y^2 + (z \pm \frac{1}{2})^2 = \frac{1}{4}$ using the relation $z = r\cos\theta$. This state is not surprisingly referred to as a $p_z$ orbital. The d state ($l = 2$, $m = 0$) has positive lobes along the $z$ axis with a negative lobe in the equatorial $xy$ plane. This state is referred to as a $d_{3z^2-r^2}$ orbital, or simply $d_{z^2}$. We will see later that the signs of the lobes are very important when considering the overlap of orbitals during bonding. It is clear from Fig. 2.14 that the sign of the s and d states are unchanged on inversion about the origin (i.e. they are even or *gerade*) whereas the p state changes sign (i.e. it is odd or *ungerade*).

Figure 2.15 shows the corresponding probability clouds $|Y_l^m(\theta, \phi)|^2$ that give the angular dependence of the probability density $|\psi_{nlm}(\mathbf{r})|^2$. Comparing with the previous figure we see that the s state remains spherical whereas the lobes of the $p_z$ state are distorted from sphericity along the $z$ axis, reflecting the change from $\cos\theta$ to $\cos^2\theta$. Since we often deal with atoms in a cubic environment, it is customary to form $p_x$ and $p_y$ orbitals by taking linear combinations of the two remaining $l = 1$ states corresponding to $m = \pm 1$. They are illustrated in Fig. 2.15(b). Note that a full p shell has spherical symmetry since $x^2 + y^2 + z^2 = r^2$. The probability clouds of the five d orbitals corresponding to $l = 2$ are shown in Fig. 2.15(c). As might

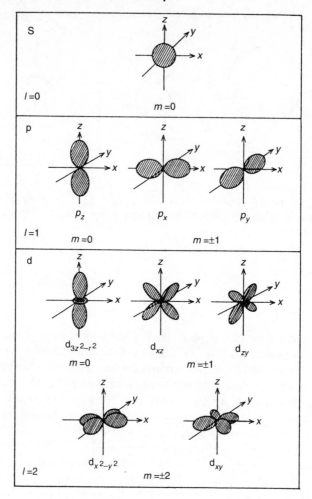

**Fig. 2.15** The probability clouds corresponding to s, p and d orbitals are shown in (a), (b), and (c) respectively.

be expected from Fig. 2.15, we will find in the following chapters that the structure of molecules and solids is very sensitive to the angular character of the valence orbitals.

The energy levels of the hydrogen atom are determined solely by the principal quantum number, $n$, through $E_n = -n^{-2}$ Rydbergs. In general, however, states with the same principal quantum number $n$, but different orbital quantum numbers, $l$, have their degeneracy lifted because the presence of more than one electron outside the nucleus leads to the potential $V(r)$ no longer showing the simple inverse distance-dependence of the hydrogenic case. This is illustrated in Fig. 2.16 where it is clear, for example, that

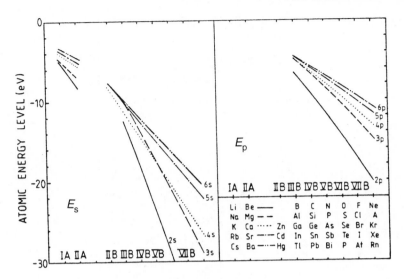

**Fig. 2.16** The valence s and p energy levels (after Herman and Skillman (1963)).

the valence 2s level of the second-row elements B to Ne lies well below that of the corresponding valence 2p level. These atomic energy levels of the *valence* electrons were taken from the tables compiled by Herman and Skillman (1963) who solved the Schrödinger equation numerically for all elements in the periodic table.

Figure 2.16 illustrates several important features to which we will be returning throughout this book.

1. The valence energy levels vary linearly across a given period just like the inverse radii in Fig. 2.13. As the nuclear charge $Ze$ increases, the electrons are bound more tightly to the nucleus. However, rather than varying as $Z^2$, which would be the result for the energy levels of a hydrogenic ion of charge $Ze$, the presence of the other valence electrons induces the linear behaviour observed.

2. The valence s and p energy levels become less strongly bound as one moves down a given group, and the value of the principal quantum number increases. This is to be expected from the $1/n^2$ dependence of the hydrogenic energy levels. But there is an exception to this rule: the 4s valence level has come down and crosses below the 3s valence level to the left of group VB. This is a direct consequence of the presence of the occupied 3d shell whose electrons do not completely screen the core from the valence 4s electrons, which therefore feel a more attractive potential than their 3s counterparts in the preceding row. This reversal in the expected ordering of the valence s energy levels is reflected in the ordering of the string between

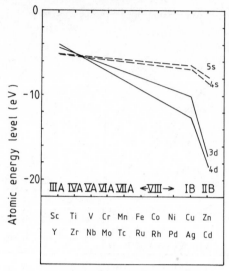

**Fig. 2.17** The valence s and d energy levels across the 3d and 4d transition metal series (after Herman and Skillman (1963)).

Al and Ga in Fig. 1.8, an anomalous ordering that also reflects their relative core sizes as we saw earlier.

3. It is clear from Fig. 2.16 that the energy difference $E_p - E_s$ decreases as one goes from right to left across a given period. This will strongly influence the nature of the energy bands and the bonding in the bulk, since if the energy difference is small, s and p electrons will hybridize to form common sp bands.

Figure 2.17 shows the valence s and d energy levels across the 3d and 4d transition metal series. The energy levels correspond to the atomic con-figuration $d^{N-1}s$, where $N$ is the total number of valence electrons, because this is the configuration closest to that of the bulk metal. Again there are several important features.

1. We see that the energy variation is linear across the transition metal series as the d shell is progressively filled with electrons. However, once the noble metal group IB is reached, the d shell contains its full complement of ten electrons, so that any further increase in atomic number $Z$ adds the additional valence electrons to the sp outer shell and pulls the d energy rapidly down, as is evidenced by the change of slope in Fig. 2.17.

2. Whereas the valence s energy level becomes slightly less strongly bound as one moves down a given group, the valence 4d energy level becomes more strongly bound than the valence 3d away from the beginning of the transition-metal series. This behaviour appears to be related to the mutual coulomb repulsion between the negatively charged valence electrons. The

3d orbitals are much more compact than the 4d orbitals, so that the putting of electrons into the 3d shell leads to a more rapid increase in repulsive energy than in the 4d shell. This leads to an *increase* in the relative separation of the valence s and d energy levels in going down a column from 3d to 4d away from the left-hand side of the series. For example, $E_s - E_d$ is about 3 eV in Cu but 6 eV in Ag, which is reflected in their different elemental colours. Relativistic effects, however, which are not included in the Schrödinger equation (2.49), reverse this trend on proceeding down to the 5d row. Unlike p and d orbitals, s orbitals have weight at the origin (cf Fig. 2.12), so they experience the singularity in the unscreened nuclear potential. For the large atomic numbers of the 5d transition elements this is sufficiently strong to accelerate the s electrons to relativistic speeds and to lower the energy of the valence 6s orbital by several electron volts, thereby *decreasing* the sd separation. This change in the relative s to d stability on moving down a given column is demonstrated by Ni, Pd and Pt whose most stable free atomic configuration changes from $3d^8 4s^2$ to $4d^{10}$ to $5d^9 6s$ respectively. This is reflected by the ordering of the string within this column in Fig. 1.8.

3. We see from the values of the s–d separation in Fig. 2.17 that we expect the bonding and structural influence of the d states to be much more marked for the divalent alkaline earths Ca and Sr at the beginning of the transition series than for the divalent elements Zn and Cd at the end. Thus it is not unreasonable that purely sp-valent Be and Mg are found to be grouped in Fig. 1.8 with Zn, Cd and Hg rather than Ca, Sr and Ba.

## 2.7. Quantum mechanical structural predictions

The trends in the free atomic energy levels and core sizes have helped us rationalize *a posteriori* some of the observed anomalies in the running of the string through the periodic table in Fig. 1.8. However, the *a priori* prediction of which crystal structure is most stable for any particular element or compound requires a careful comparison of the total energies of many different competing structure types. At the outset this appears a formidable task. The Schrödinger equation can be solved exactly for the hydrogen atom. Its solution for the hydrogen molecule and for all other atoms is a many-body problem. The wave function is no longer dependent on the coordinates of a single electron but on the coordinates of all $N$ electrons that are present. For a thimble full of bulk material, $N$ would typically be of the order of $10^{23}$.

The way forward has already been hinted at in our discussion of the electron gas. The simplest approximation due to Hartree in 1928 is to assume that the individual electrons move independently of each other, so that each electron feels the average electrostatic field of all the other electrons in addition to the potential from the ionic lattice. This average field has to be determined self-consistently in that the input charge density, which enters the electrostatic potential on the left-hand side of the Schrödinger equation, depends on the probability density or wave function of the individual electrons that is predicted as output. Alas, just as for the simplest case of jellium, the independent particle approximation fails to describe bonding. The bond in aluminium, for example, is predicted to be more than two orders of magnitude smaller than that observed experimentally.

In 1930 Fock extended the theory by including exchange which lowered the total binding energy by keeping parallel spin electrons apart through Pauli's exclusion principle. The so-called Hartree–Fock approximation, however, still made a sizeable error because it neglected correlations in the motion between anti-parallel spin electrons. This error leads to the electronic heat capacity of metals varying with temperature as $T/\log T$ whereas experimentally we have already seen it displays a simple linear temperature-dependence. But to go beyond the Hartree–Fock approximation seemed very hard except for the simplest system of the free-electron gas. Even quite recently, Hume-Rothery (1962) was stressing in the preface of his textbook *Atomic Theory for Students of Metallurgy* 'the extreme difficulty of producing any really quantitative electron theory'.

The breakthrough came two years later when Hohenberg, Kohn and Sham proved that the *total ground-state energy* of a many-electron system is a functional of the density (Hohenberg and Kohn 1964; Kohn and Sham 1965). (A functional is a function of a function, the electron density being the function $\rho(r)$.) This seemingly simple result, by focusing on the electron density rather than the many-body wave function, allowed them to derive an effective one-electron-type Schrödinger equation, namely

$$-\frac{\hbar^2}{2m}\nabla^2\psi(\mathbf{r}) + [V_H(\mathbf{r}) + V_{xc}(\mathbf{r})]\psi(\mathbf{r}) = E\psi(\mathbf{r}). \qquad (2.61)$$

It is directly analogous to Hartree's, except that in addition to the average electrostatic potential $V_H(\mathbf{r})$, each electron also feels a further attractive potential, the so-called exchange-correlation potential $V_{xc}(\mathbf{r})$. Just as we saw earlier for the free-electron gas this exchange-correlation potential arises from each electron being surrounded by its own mutual exclusion zone or hole from which other electrons are kept out (cf Fig. 2.10). In practice, the exact shape of this exchange-correlation hole is not known except for the homogeneous free-electron gas. Hohenberg, Kohn and Sham, therefore, replaced the exact hole by the hole which an electron would have in a free-electron gas with the same density as that seen locally by the given electron at any particular instant—the so-called local density approximation (LDA).

Although initially it was thought that this approximation would only work well for those systems with nearly uniform or homogeneous electron densities, in practice, extensive computations have demonstrated the surprising accuracy of LDA in predicting the structural properties of a wide

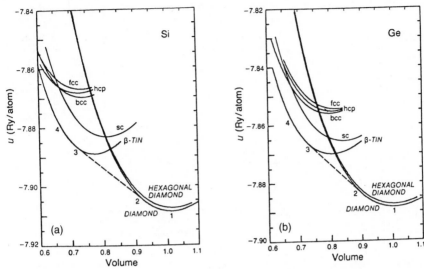

**Fig. 2.18** The binding energy curves of (a) Si and (b) Ge for seven different crystal structures. The volume has been normalized by the equilibrium atomic volume. The dashed line is the common tangent of the energy curves for the diamond and $\beta$-tin phase, the system moving from $1 \to 2 \to 3 \to 4$ under pressure. (From Yin and Cohen (1982).)

range of ionic, covalent and metallic materials. This accuracy of prediction is due to the robustness of the concept of the exchange-correlation hole— even though its *shape* might be poorly represented by LDA, the attractive potential which the electron sees at its *centre* is well described as the approximate and exact holes both exclude precisely one electron each.

We illustrate the accuracy of the LDA in making structural predictions with two examples. The first example compares the behaviour of covalently bonded Si and Ge under pressure. We see from Fig. 2.18 that LDA predicts that the semiconducting diamond cubic phase transforms to the metallic $\beta$-Sn phase under pressure as observed experimentally. We also find the lowering in the stability of the close-packed structure types fcc, bcc, and hcp with respect to the diamond lattice on going from Si to Ge that reflects the known trend down group IV in Table 1.1. The second example compares the structural behaviour of the four neighbouring intermetallics ScAl, TiAl, YAl and ZrAl. We see from Fig. 2.19 that LDA predicts that from amongst the nine structure types considered, ScAl takes the ground-state structure cP2(CsCl) 14/14, TiAl takes tP4(CuAu) 12/12, and YAl and ZrAl take oS8(CrB) 17/9 as observed within the AB structure map at the end of the book.

Thus, we have come a long way from the exactly soluble problems of quantum mechanics, the free-electron gas and the hydrogen atom. The concept of the exchange-correlation hole linked with the LDA has allowed

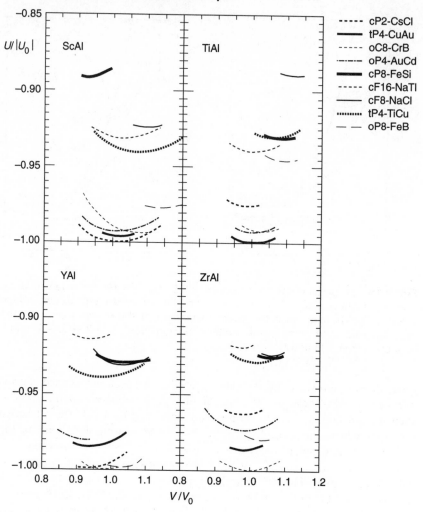

**Fig. 2.19** The binding energy curves of ScAl, TiAl, YAl, and ZrAl for nine different structure types. The energy and volumes have been normalized by the respective predicted equilibrium values of the ground-state structure. (After Nguyen Manh *et al.* (1995).)

reliable numerical predictions to be made on the bonding and structure of ionic, covalent and metallic systems. Moreover, it has placed the many-body quantum mechanical problem within an effective one-electron framework so beloved of chemists in their molecular orbital description of molecules and physicists in their band theory description of solids. We pursue these one electron ideas further in the following chapters.

# References

Douglas, B. E., McDaniel, D. H., and Alexander, J. J. (1983). *Concepts and models of inorganic chemistry*. Wiley, New York.

Gasiorowicz, S. (1974). *Quantum physics*. Wiley, New York.

Herman, F. and Skillman, S. (1963). *Atomic structure calculations*. Prentice-Hall, Englewood Cliffs, New Jersey.

Hohenberg, P. and Kohn, W. (1964). *Physical Review* **136**, B864.

Hume-Rothery, W. (1962). *Atomic theory for students of metallurgy* (Fourth revised reprint). Institute of Metals, London.

Kohn, W. and Sham, L. J. (1965). *Physical Review* **140**, A1133.

Nguyen Manh, D., Bratkovsky, A. M., and Pettifor, D. G. (1995). *Philosophical Transactions of the Royal Society of London* **351**, 529.

Yin, M. T. and Cohen, M. L. (1982). *Physical Review* B **26**, 5668.

Zunger, A. (1980). *Physical Review* B **22**, 5849.

# 3
# Bonding of molecules

## 3.1. Introduction

The concepts which we need for understanding the structural trends within covalently bonded solids are most easily introduced by first considering the much simpler system of diatomic molecules. They are well described within the molecular orbital (MO) framework that is based on the overlapping of atomic wave functions. This picture, therefore, makes direct contact with the properties of the individual free atoms which we discussed in the previous chapter, in particular the atomic energy levels and angular character of the valence orbitals. We will see that ubiquitous quantum mechanical concepts such as the covalent bond, overlap repulsion, hybrid orbitals, and the relative degree of covalency versus ionicity all arise naturally from solutions of the one-electron Schrödinger equation for diatomic molecules such as $H_2$, $N_2$, and LiH.

## 3.2. Bond formation in s-valent dimers

Let us consider what happens as two s-valent atoms A and B are brought together from infinity to form the AB diatomic molecule as illustrated schematically in Fig. 3.1. The more deeply bound energy level $E_A$ could represent, for example, the hydrogenic 1s orbital ($E_A = -13.6$ eV), whereas the less deeply bound energy level $E_B$ could represent lithium's 2s orbital ($E_B = -5.5$ eV. cf Fig. 2.16). Each free atomic orbital satisfies its own effective one-electron Schrödinger equation (cf eqn (2.49)), namely

$$-\frac{\hbar^2}{2m}\nabla^2\psi_A(\mathbf{r}) + V_A(r)\psi_A(\mathbf{r}) = E_A\psi_A(\mathbf{r}) \tag{3.1}$$

and

$$-\frac{\hbar^2}{2m}\nabla^2\psi_B(\mathbf{r}) + V_B(r)\psi_B(\mathbf{r}) = E_B\psi_B(\mathbf{r}), \tag{3.2}$$

where $V(r)$ is the total potential seen by the electron, including both the Hartree and exchange-correlation contributions.

We wish to solve the one electron Schrödinger equation for the AB dimer,

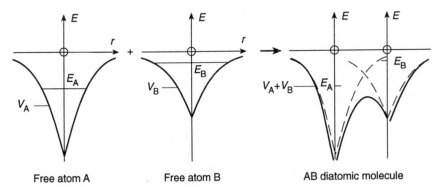

**Fig. 3.1** The potential of the AB diatomic molecule may be approximated by the overlapping of the free atom A and B potentials. $E_A$ and $E_B$ are the valence s electron energy levels.

namely

$$-\frac{\hbar^2}{2m}\nabla^2\psi_{AB}(\mathbf{r}) + V_{AB}(r)\psi_{AB}(\mathbf{r}) = E\psi_{AB}(\mathbf{r}) \tag{3.3}$$

where $V_{AB}(\mathbf{r})$ is the molecular potential. In principle, this potential should be determined self-consistently in that both the Hartree and exchange-correlation potentials which enter as *input* on the left-hand side of eqn (3.3) depend on the average electronic charge density, and hence on the *output* electronic wave function, $\psi_{AB}$. In practice we will approximate this self-consistent molecular potential by the sum of the individual free atomic potentials, so that

$$V_{AB} = V_A + V_B, \tag{3.4}$$

as sketched in Fig. 3.1. This is a good approximation for the covalently bonded systems that are of prime interest in this book. However, it is a poorer approximation for systems towards the ionic end of the bonding spectrum where explicit shifts in the energy levels due to the flow of electrons from one atom to the other must be incorporated, as is done later in our treatment of the heats of formation of transition metal alloys in §7.6.

The molecular Schrödinger equation can be solved exactly for the case of $H_2^+$ when $V_{AB}$ is simply the sum of two hydrogen ion potentials. In general, however, an exact solution is not possible. Following the well-worn tracks of MO theory we look instead for an approximate solution that is given by some linear combination of atomic orbitals (LCAO). Considering the AB dimer illustrated in Fig. 3.1 we write

$$\psi_{AB} = c_A\psi_A + c_B\psi_B \tag{3.5}$$

where $c_A$ and $c_B$ are two constant coefficients to be determined. It then follows from the Schrödinger equation (3.3) that

$$(\hat{H} - E)(c_A\psi_A + c_B\psi_B) = 0, \tag{3.6}$$

## 52    Bonding of molecules

where $\hat{H}$ is the Hamiltonian operator for the AB dimer, namely

$$\hat{H} = -\frac{\hbar^2}{2m}\nabla^2 + V_{AB}.$$

Premultiplying by $\psi_A$ (or $\psi_B$) and integrating over all space we find the LCAO secular equation (taking $\psi^* = \psi$ as $\psi$ is real for s orbitals)

$$\begin{bmatrix} H_{AA} - E & H_{AB} - ES_{AB} \\ H_{BA} - ES_{BA} & H_{BB} - E \end{bmatrix}\begin{bmatrix} c_A \\ c_B \end{bmatrix} = 0, \tag{3.7}$$

where the Hamiltonian and overlap matrix elements are given by

$$H_{\alpha\beta} = \int \psi_\alpha \hat{H}\psi_\beta \, d\mathbf{r} \tag{3.8}$$

and

$$S_{\alpha\beta} = \int \psi_\alpha \psi_\beta \, d\mathbf{r}. \tag{3.9}$$

The Hamiltonian matrix elements can be evaluated directly by assuming $V_{AB} = V_A + V_B$ as in eqn (3.4). The *diagonal* element $H_{AA}$ may then be written

$$H_{AA} = \int \psi_A(\hat{H}_A + V_B)\psi_A \, d\mathbf{r} = \int \psi_A(E_A + V_B)\psi_A \, d\mathbf{r} \tag{3.10}$$

where $\hat{H}_A = -(\hbar^2/2m)\nabla^2 + V_A$ is the Hamiltonian operator for the free atom, A, and the last identity follows from the Schrödinger equation (3.1).
Thus,

$$H_{AA} = E_A + \int \rho_A V_B \, d\mathbf{r}, \tag{3.11}$$

where $\rho_A = \psi_A^2$ is the electronic probability or number density of free atom, A. The diagonal element, $H_{AA}$, is, therefore, given by the free atomic energy level, $E_A$, shifted by the so-called crystal field term, which reflects the lowering in energy of the electron on atom A due to the attractive tail of the potential from the neighbouring atom, B (cf the right-hand sketch in Fig. 3.1). We shall neglect the crystal field in the following discussion as it does not affect the fundamental description of the covalent bond. We shall, therefore, approximate $H_{AA}$ and $H_{BB}$ by the free atomic energy levels $E_A$ and $E_B$ respectively.
The *off-diagonal* element $H_{AB}$ may be written as

$$H_{AB} = \int \psi_A(\hat{H}_A + V_B)\psi_B \, d\mathbf{r} = \int \psi_A(E_A + V_B)\psi_B \, d\mathbf{r}, \tag{3.12}$$

where we have used the fact that the Hamiltonian operator, $\hat{H}_A$, is hermitian to act with it to the left on $\psi_A$ under the first integral sign. However, we could equally well have written $H_{AB}$ as

$$H_{AB} = \int \psi_A(\hat{H}_B + V_A)\psi_B \, dr = \int \psi_A(E_B + V_A)\psi_B \, dr, \tag{3.13}$$

where $\hat{H}_B = -(\hbar^2/2m)\nabla^2 + V_B$ is the Hamiltonian operator for the free atom, B.

Thus, treating the A and B atoms on an equal footing by adding equations (3.12) and (3.13), we have that

$$H_{AB} = \int \psi_A(\bar{V} + \bar{E})\psi_B \, dr, \tag{3.14}$$

where $\bar{V} = \frac{1}{2}(V_A + V_B)$ and $\bar{E} = \frac{1}{2}(E_A + E_B)$ are the average values of the atomic potentials and energy levels respectively. Finally, therefore, the off-diagonal element is given by

$$H_{AB} = h + \bar{E}S, \tag{3.15}$$

where $h = \int \psi_A \bar{V}\psi_B \, dr$ and $S = \int \psi_A \psi_B \, dr = S_{AB}$ are the *bond* and *overlap* integrals respectively. The bond integral is negative, since the two positive s orbitals are overlapping in the negative molecular potential of Fig. 3.1.

The LCAO secular equation then takes the form

$$\begin{bmatrix} -\frac{1}{2}\Delta E - (E - \bar{E}) & h - (E - \bar{E})S \\ h - (E - \bar{E})S & \frac{1}{2}\Delta E - (E - \bar{E}) \end{bmatrix} \begin{bmatrix} c_A \\ c_B \end{bmatrix} = 0, \tag{3.16}$$

where $\Delta E = E_B - E_A$ is the atomic energy level mismatch (cf $E_A = \bar{E} - \frac{1}{2}\Delta E$, $E_B = \bar{E} + \frac{1}{2}\Delta E$). Unlike eqn (3.7) we see that the eigenvalues, $E$, are now measured explicitly with respect to the average energy, $\bar{E}$. Equation (3.16) has non-trivial solutions if the secular determinant vanishes, that is

$$\begin{vmatrix} -\frac{1}{2}\Delta E - (E - \bar{E}) & h - (E - \bar{E})S \\ h - (E - \bar{E})S & \frac{1}{2}\Delta E - (E - \bar{E}) \end{vmatrix} = 0. \tag{3.17}$$

The resultant quadratic equation may be solved directly to yield

$$E^{\pm} = \bar{E} + \{|h|S \mp \frac{1}{2}[4h^2 + (1 - S^2)(\Delta E)^2]^{1/2}\}/(1 - S^2). \tag{3.18}$$

As we are interested in extracting the essential ingredients that characterize bond formation, we may neglect second-order terms in the overlap integral, $S$. By expanding $(1 - S^2)^{-1}$ as $1 + S^2 + \cdots$ we see from eqn (3.18) that these $S^2$ terms lead to eigenvalue shifts such as $hS^2$ which are *third*-order or higher in the small quantities $h$ and $S$ that vanish as the bond is pulled apart. The eigenvalues may, therefore, be approximated by

$$E^{\pm} = \bar{E} + |h|S \mp \frac{1}{2}[4h^2 + (\Delta E)^2]^{1/2}. \tag{3.19}$$

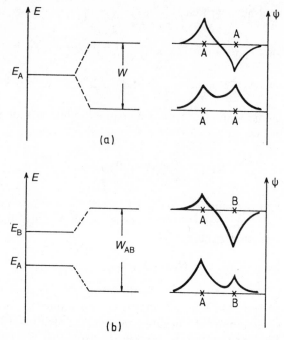

**Fig. 3.2** The bonding and antibonding states for (a) the homonuclear and (b) the heteronuclear diatomic molecule. The shift in the energy levels due to overlap repulsion has not been shown.

Therefore, as illustrated by the left-hand panels of Fig. 3.2, s-valent diatomic molecules are characterized by bonding and antibonding states that are separated in energy by the amount $w_{AB}$, such that

$$w_{AB}^2 = 4h^2 + (\Delta E)^2. \qquad (3.20)$$

It is now clear why $h$ is referred to as the bond integral, since for homonuclear diatomic molecules the bonding state has been shifted downwards by the amount $|h|$. We also see from eqn (3.19) that the molecular eigenvalues are shifted upwards by the amount $|h|S$ due to overlap repulsion that reflects Pauli's exclusion principle. These are the two key ingredients of the covalent bond: an attractive bond energy pulling the atoms together, which is balanced at equilibrium by a repulsive overlap potential keeping the atoms apart.

It follows from the LCAO secular equation (3.16) that the eigenfunctions corresponding to the eigenvalues $E^{\pm}$ are given to first order in $S$ by

$$\psi_{AB}^{\pm} = c_A^{\pm}\psi_A + c_B^{\pm}\psi_B, \qquad (3.21)$$

where

$$c_A^{\pm} = \frac{1}{\sqrt{2}}\left[1 \pm \frac{\delta - S}{\sqrt{1 + \delta^2}}\right]^{1/2} \tag{3.22}$$

and

$$c_B^{\pm} = \pm\frac{1}{\sqrt{2}}\left[1 \mp \frac{\delta + S}{\sqrt{1 + \delta^2}}\right]^{1/2} \tag{3.23}$$

with $\delta = \Delta E/2|h|$ being the normalized atomic energy-level mismatch. Thus, as shown by the upper right-hand panel of Fig. 3.2, the bonding and antibonding states of a homonuclear diatomic molecule are given by the symmetric and antisymmetric combinations of the atomic orbitals $\psi_A$ and $\psi_B$ respectively. (Note that from eqs (3.22) and (3.23)

$$c_A^{\pm} = \pm c_B^{\pm} = \frac{1}{\sqrt{2}}(1 \mp S)^{1/2}$$

which is equivalent to first order in $S$ to the usual normalization prefactor of hydrogenic molecular orbitals, namely $(1/\sqrt{2})(1 \pm S)^{-1/2}$. As expected, the lower right-hand panel of Fig. 3.2 shows that the bonding electrons in a heteronuclear diatomic molecule spend more time on the more attractive site than on the less attractive site.

We see that the formation of the bond is accompanied by a marked redistribution of the electronic charge. Since the bonding state $\psi_{AB}^+$ is occupied by two valence electrons of opposite spin, the electronic probability or number density of the diatomic molecule will be given by $\rho_{AB}(\mathbf{r}) = 2[\psi_{AB}^+(\mathbf{r})]^2$ with the corresponding electronic charge density $-e\rho_{AB}(\mathbf{r})$ where $e$ is the magnitude of the electronic charge. From eqs (3.21)–(3.23), the electronic density may be written in the form (neglecting second-order terms in the overlap $S$)

$$\rho_{AB}(\mathbf{r}) = (1 + \alpha_i)\rho_A(\mathbf{r}) + (1 - \alpha_i)\rho_B(\mathbf{r}) + \alpha_c\rho_{bond}(\mathbf{r}), \tag{3.24}$$

where

$$\rho_{A(B)}(\mathbf{r}) = [\psi_{A(B)}(\mathbf{r})]^2 \tag{3.25}$$

and

$$\rho_{bond}(\mathbf{r}) = 2\psi_A(\mathbf{r})\psi_B(\mathbf{r}) - S[\rho_A(\mathbf{r}) + \rho_B(\mathbf{r})]. \tag{3.26}$$

The terms $\alpha_i$ and $\alpha_c$ are determined by the normalized energy-level mismatch $\delta$ through

$$\alpha_i = \delta/(1 + \delta^2)^{1/2} \tag{3.27}$$

and

$$\alpha_c = 1/(1 + \delta^2)^{1/2}. \tag{3.28}$$

For *homonuclear* diatomic molecules the atomic energy-level mismatch vanishes so that $\alpha_i = 0$ and $\alpha_c = 1$. Hence, the change in the electronic

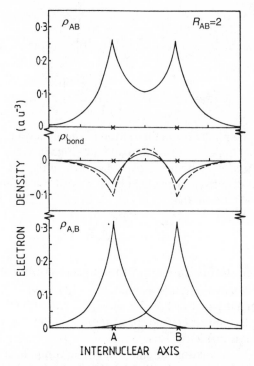

**Fig. 3.3** The electron density of the homonuclear molecule (upper panel) can be regarded as the sum of the non-interacting or frozen free-atom electron densities (lower panel) and the quantum mechanically induced bond density (middle panel). The dashed curve represents the first-order approximation, eqn (3.26), for the bond density, the deviation from the exact result (full curve) being due to the sizeable value of the overlap integral namely $S = 0.59$ at $R = 2$ au.

charge distribution on forming the molecule is given solely by the bond charge contribution $-e\rho_{bond}$. This is illustrated in Fig. 3.3 for the case of the hydrogen molecule where we see that, as expected, the electronic charge has moved from the outer region of the molecule into the strongly attractive bond region between the atoms. This is a truly quantum mechanical effect, since it reflects the constructive interference between the wave functions $\psi_A(\mathbf{r})$ and $\psi_B(\mathbf{r})$ that are centred on the two atomic sites. We should note that the total charge associated with $\rho_{bond}$ over all space is identically zero since from eqs (3.9) and (3.26)

$$\int \rho_{bond}(\mathbf{r})\, d\mathbf{r} = 2S - 2S = 0. \tag{3.29}$$

For *heteronuclear* diatomic molecules, the atomic energy-level mismatch does not vanish, so that $\delta \neq 0$. Hence, the electronic charge distribution

$-e\rho_{AB}(\mathbf{r})$ contains the ionic contributions $-e\alpha_i\rho_A(\mathbf{r})$ and $+e\alpha_i\rho_B(\mathbf{r})$ in addition to the covalent bond charge contribution $-e\alpha_c\rho_{bond}(\mathbf{r})$. Thus the amplitude of the ionic charge is proportional to $\alpha_i$, whereas the amplitude of the covalent bond charge is proportional to $\alpha_c$. This has important implications: the strongest possible covalent bond between isovalent atoms occurs when the valence energy levels on the two sites shows zero mismatch, so that $\delta = 0$, and $\alpha_c$ takes its maximum allowed value of unity. As the energy-level mismatch increases the electron spends more time on a given site rather than being shared equally between the two sites in the covalent bond region. This is illustrated by the very different behaviour of the homonuclear and heteronuclear bonding state eigenfunctions in Fig. 3.2.

The terms $\alpha_i$ and $\alpha_c$ are said to measure the degree of ionicity and covalency of the bond. From eqs (3.27) and (3.28) they satisfy the constraint

$$\alpha_i^2 + \alpha_c^2 = 1. \tag{3.30}$$

For a homonuclear diatomic molecule, the bond is purely covalent ($\alpha_i = 0$, $\alpha_c = 1$) whereas for a heteronuclear diatomic molecule the bond shows mixed covalent-ionic character ($\alpha_i \neq 0$, $\alpha_c \neq 1$). In the limit as the separation between the atomic energy levels on the two atoms becomes very large the bond becomes purely ionic ($\alpha_i = 1$, $\alpha_c = 0$).

## 3.3. Electronegativity scales

The degree of ionicity of an *s-valent* bond can be written explicitly from eqn (3.27) in terms of the bond integral $h$ and atomic energy-level mismatch $\Delta E$ as

$$\alpha_i = \Delta E/[4h^2 + (\Delta E)^2]^{1/2}. \tag{3.31}$$

A not unrelated definition of $\alpha_i$ for *sp-valent* octet AB compounds has been given by Phillips and Van Vechten (1969). These octet compounds such as NaCl and ZnS have eight valence s and p electrons per AB unit. Phillips and Van Vechten assumed that the average energy gap $E_g$ of these semiconducting or insulating compounds is made up of covalent and ionic contributions, $E_c$ and $E_i$ respectively, that are related via eqn (3.20) by

$$E_g^2 = E_c^2 + E_i^2. \tag{3.32}$$

The average energy gap or the splitting between the bonding and antibonding states may be loosely thought of as the energy difference between the centres of gravity of their valence and conduction bands, which was determined using photoemission spectroscopy. Thus, by measuring $E_c$ for the elemental group IV semiconductors and $E_g$ for the octet compound semiconductors or insulators, the appropriate values of $E_i$ could be deduced from eqn (3.32). For example, the isoelectronic series of compounds Ge, GaAs, ZnSe, and CuBr were found to take values of the ionic component $E_i$ equal to 0.0, 1.9, 3.8, and 5.6 eV respectively. This linear increase across the isoelectronic series is not unexpected from the linear behaviour of the 4s and 4p atomic energy levels that are shown in Fig. 2.16. These values of $E_i$ should be compared to the value of the covalent component for this isoelectronic series, namely $E_c = 5.6$ eV (the value for elemental Ge).

The degree of ionicity of the bond in these octet compounds is then defined by analogy with eqn (3.31) as

$$\alpha_i = E_i/[E_c^2 + E_i^2]^{1/2}. \qquad (3.33)$$

Thus as we go across the isoelectronic series Ge → GaAs → ZnSe → CuBr, the degree of ionicity increases from 0 → 0.32 → 0.56 → 0.71 as expected. A noteable achievement of the Phillips–Van Vechten ionicity scale was that it also allowed an excellent structural ordering of the sp-valent octet compounds. As shown in Fig. 3.4, the structure map $(E_c, E_i)$ separates all those

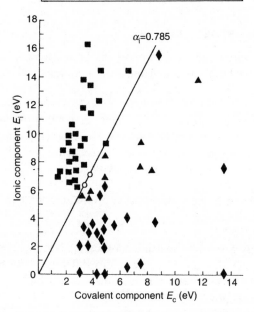

**Fig. 3.4** The Phillips–Van Vechten structure map $(E_c, E_i)$ for the sp-valent octet AB compounds. The four-fold coordinated zinc blende and wurtzite structure types are separated from the six-fold coordinated NaCl structure type by the straight line corresponding to the degree of ionicity $\alpha_i = 0.785$. (After Phillips and Van Vechten (1969).)

compounds with the four-fold zinc blende or wurtzite structure types from those with the six-fold NaCl structure type. In fact, the boundary is the straight line given by $E_i = 1.27E_c$ that corresponds to the degree of ionicity $\alpha_i = 0.785$. All those sp-valent octet compounds with ionicities less than this critical value are tetrahedrally coordinated; all those with ionicities greater than this are octahedrally coordinated. This is consistent with conventional wisdom that zinc blende and wurtzite are covalent structure types, whereas NaCl is an ionic structure type.

We have seen from eqn (3.24) that the degree of ionicity $\alpha_i$ reflects the relative ability of two different atoms to attract electrons to themselves. It is, therefore, a direct measure of their electronegativity difference $\Delta X$, where $X$ is the electronegativity of the individual atom. We could, therefore, set up two *restricted* electronegativity scales, the one for s-valent atoms that implies $\Delta X \alpha \Delta E$ (assuming that $h$ is approximately constant), and the other for sp-valent atoms in tetrahedral or octahedral environments that implies $\Delta X \alpha E_i$ (assuming that $E_c$ is approximately constant). It is more common in practice, however, to work with *general* electronegativity scales that can in principle be applied to all atoms within the periodic table.

Figure 3.5 shows the two most widely used scales according to Pauling (1960) and Mulliken (1934) respectively. Pauling based his scale on the known experimental heats of formation $\Delta H$ of binary molecules and compounds. Assuming that the excess AB bond energy was ionic in origin, he defined $\Delta X$ through (in its simplest, earliest form)

$$\Delta H = U_{AB} - \tfrac{1}{2}(U_{AA} + U_{BB}) = -k(\Delta X)^2 \qquad (3.34)$$

where $k$ is a constant and $U_{AA}$, $U_{BB}$ and $U_{AB}$ are the dissociation or binding energies of the $A_2$, $B_2$ and AB diatomic molecules respectively. Mulliken, on the other hand, based his scale on the known ionization potentials $I$ and electron affinities $A$ of the individual free atoms, defining

$$X = \tfrac{1}{2}(I + A) \qquad (3.35)$$

This was justified as follows. The energy required to take an electron from a neutral atom Y to a neutral atom Z is $I_Y - A_Z$, whereas the energy cost to take an electron from a neutral atom Z to a neutral atom Y is $I_Z - A_Y$. Hence, the two atoms Y and Z would have an equal propensity for attracting electrons or equal electronegativity if $I_Y - A_Z = I_Z - A_Y$, that is if $I_Y + A_Y = I_Z + A_Z$. This is consistent with the Mulliken definition, eqn (3.35), the factor $1/2$ being arbitrary. As can be seen by comparing the two different scales in Fig. 3.5, the Mulliken values are approximately 2.8 times the Pauling values.

The concept of electronegativity is easy to visualize but difficult to apply quantitatively. Mulliken's definition, based on free-atomic properties, does not account for the influence of the local atomic environment on the ionic state that is reflected, for example, in the fact that the degree of ionicity $\alpha_i$

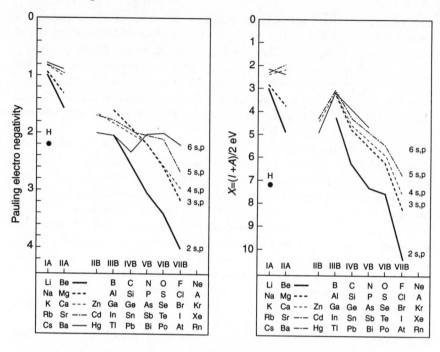

**Fig. 3.5** The Pauling (left-hand panel) and Mulliken (right-hand panel) electro-negativities for the sp-valent elements. Note that the electronegativities scales run vertically *downwards* in order to emphasize the similarity with the free-atom energy levels in Fig. 2.16 and negative inverse core sizes in Fig. 2.13.

depends on both the atomic energy-level mismatch, $\Delta E$, and the bond integral, $h$. Pauling's definition, based on the heat of formation, implies that all binary compounds form since $-(\Delta X)^2$ is negative (or at most zero). However, we have already seen in Fig. 1.9 that nearly three-quarters of all possible binary AB compounds do *not* form. The problem is that electronegativity is essentially a classical concept with electrons flowing from electropositive to electronegative atoms, thereby setting up an attractive ionic bond. In practice, as we shall demonstrate in future chapters, the bonding and structure of most binary systems are controlled by quantum mechanical rather than classical contributions to the energy.

## 3.4. Dissociation of the hydrogen dimer

In the early days of quantum mechanics, MO theory received a very bad press compared to valence bond (VB) theory because it predicted the incorrect dissociation behaviour of the hydrogen molecule. This is illustrated

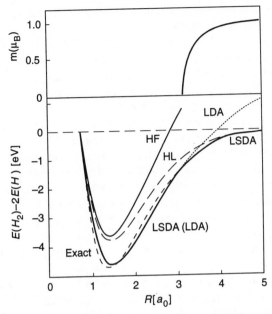

**Fig. 3.6** Binding energy curves for the hydrogen molecule (lower panel). HF and HL are the Hartree–Fock and Heitler–London predictions, whereas LDA and LSDA are those for local density and local spin density approximations respectively. The upper panel gives the local magnetic moment within the LSDA self-consistent calculations. (After Gunnarsson and Lundquist (1976).)

by Fig. 3.6. We see that a variational Heitler–London VB solution (to be discussed below) gives not only about 80% of the observed binding energy, but also dissociates correctly as the atoms are pulled apart to infinity. A variational Hartree–Fock MO solution, on the other hand, not only provides slightly less binding energy at equilibrium, but more seriously dissociates to the wrong limit.

The origin of this discrepancy is easy to track down. The MO description of the hydrogen molecule places the two electrons with opposite spin into the bonding state $\psi_{AB}^+$, so that the two-electron wave function can be written

$$\psi(1,2) = \psi_{AB}^+(1)\psi_{AB}^+(2) = \tfrac{1}{2}[\psi_A(1) + \psi_B(1)][\psi_A(2) + \psi_B(2)], \quad (3.36)$$

where for simplicity we have neglected the overlap contribution to the normalization prefactor. Thus, multiplying out we have

$$\psi(1,2) = [\tfrac{1}{2}\psi_A(1)\psi_B(2) + \tfrac{1}{2}\psi_B(1)\psi_A(2)] + [\tfrac{1}{2}\psi_A(1)\psi_A(2) + \tfrac{1}{2}\psi_B(1)\psi_B(2)]. \quad (3.37)$$

The two terms in the first square brackets represent *covalent* configurations in which electron 1 is associated with atom A, electron 2 with atom B and vice versa. The two terms in the second square brackets represent *ionic* configurations in which both electrons 1 and 2 are associated with atom A, none with atom B and vice versa.

It is clear, therefore, that as the atoms are pulled apart the Hartree–Fock MO solution does not go over to that of two neutral-free atoms $H^0H^0$, but instead goes over to a mixed configuration, schematically represented by $\frac{1}{4}[2H^0H^0 + H^+H^- + H^-H^+]$. Since the energy cost for the ionic configuration is $I - A = (13.6 - 0.8)$ eV $= 12.8$ eV, the Hartree–Fock MO solution dissociates incorrectly to $+6.4$ eV rather than zero! (We have assumed the Hartree–Fock treatment of $H^-$ is exact.) The Heitler–London VB solution avoids this problem by working with only the covalent configurations in the first square bracket of eqn (3.37), so that it dissociates correctly.

Local Density Functional MO theory dramatically improves the predicted equilibrium binding energy compared to the Hartree–Fock solution, since it includes correlation effects explicitly through the exchange-correlation potential in eqn (3.3). However, the local density binding energy curve still dissociates incorrectly. This problem can be overcome by allowing the density functional solution to become *spin-polarized* by removing the constraint that the eigenfunctions for the up and down spin electrons must be the same. This is illustrated in Fig. 3.7(d) where we can look for an up-spin solution that is weighted on atom A, a down-spin solution on atom B. We

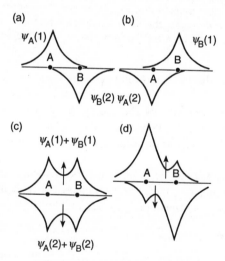

**Fig. 3.7** The Heitler–London configuration $\psi_A(1)\psi_B(2)$ and $\psi_A(2)\psi_B(1)$ (a) and (b) respectively, where $\psi_A$ and $\psi_B$ represent the atomic 1s orbitals centred on atoms A and B respectively, and 1 and 2 represent the coordinates of the two (indistinguishable) electrons. (c) The molecular orbital basis function in the singlet state where electrons 1 and 2 have opposite spin. (d) The up and down spin eigenfunctions corresponding to local exchange fields of opposite sign on A and B.

could then imagine that, as the atoms are pulled apart, the up-spin electron spends more and more of its time on atom A, the down-spin electron on atom B, so that in the limit of infinite separation we go over to the correct limit of two neutral hydrogen atom configurations.

We see from Fig. 3.6 that local spin density functional theory indeed gives an excellent binding energy curve at *all* internuclear separations. The upper panel of Fig. 3.6 shows that a spin polarized solution exists for $R > 3.2$ au, the degree of spin polarization becoming almost unity by $R = 5$ au. As might be expected by comparing Fig. 3.7(d) with Fig. 3.2(b), this onset of spin polarization as the hydrogen molecule is pulled apart can be understood by extending the simple model of the heteronuclear diatomic molecule already discussed. From eqn (2.51), a net imbalance in the number of up and down spin electrons on a given atom will correspond to a net magnetic moment $m$ on that site which for the configuration illustrated in Fig. 3.7(d) will be given by

$$m = N_A^\uparrow - N_A^\downarrow = N_B^\downarrow - N_B^\uparrow \qquad (3.38)$$

in Bohr magnetons (where $N_A^\uparrow + N_A^\downarrow = N_B^\uparrow + N_B^\downarrow = 1$). By Pauli's exclusion principle, electrons with spin parallel to the local magnetic moment will see a more attractive potential on that site than those with spin antiparallel. We expect this difference in on-site energy to be directly proportional to the magnitude of the magnetic moment, namely

$$E_A^\downarrow - E_A^\uparrow = E_B^\uparrow - E_B^\downarrow = Im \qquad (3.39)$$

where the constant of proportionality $I$ is called the Stoner *exchange integral*.

Thus, as can be seen from the left-hand panel of Fig. 3.8, the spin-polarized solution for the hydrogen molecule is equivalent to that for a

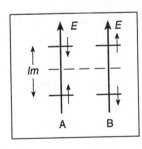
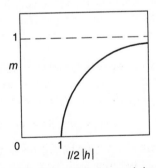

**Fig. 3.8** Left-hand panel: The on-site atomic energy levels for up and down spin electrons due to the exchange splitting $Im$ where $I$ and $m$ are the Stoner exchange integral and local moment respectively. Right-hand panel: The local magnetic moment, $m$, as a function of $I/2|h|$ where $I$ and $h$ are the exchange and bond integrals respectively. Compare with the self-consistent LSDA solution in the upper panel of Fig. 3.6.

## 64  Bonding of molecules

heteronuclear diatomic molecule with atomic energy-level mismatch

$$\Delta = E_B^\uparrow - E_A^\uparrow = E_A^\downarrow - E_B^\downarrow = Im \tag{3.40}$$

Consequently, we have from the eigensolutions, eqs (3.22) and (3.23), that

$$m = N_A^\uparrow - N_A^\downarrow = N_A^\uparrow - N_B^\uparrow = |c_A^+|^2 - |c_B^+|^2, \tag{3.41}$$

so that

$$m = \Delta/[4h^2 + \Delta^2]^{1/2}, \tag{3.42}$$

(with the overlap integral, $S$ cancelling to first order). We, therefore, have to solve for $\Delta$ self-consistently: an initial choice of the exchange splitting, $\Delta$, implies a net magnetic moment, $m$, given by eqn (3.42) which leads via Pauli's exclusion principle to an exchange splitting, $\Delta$, given by eqn (3.40).

This simple example of a self-consistent field problem has an analytic solution, since substituting eqn (3.42) into eqn (3.40), we find the exchange splitting,

$$\Delta = [I^2 - 4h^2]^{1/2}. \tag{3.43}$$

It follows that no self-consistent solution exists, unless the exchange integral, $I$, is greater than the initial bonding–antibonding energy-level separation, $2|h|$. This reflects the fact that the non-spin polarized state becomes unstable to spin polarization when the energy gain from flipping a spin (given by $-I$) overcomes the resultant increase in kinetic energy (given by $2|h|$). Finally, substituting eqn (3.43) into eqn (3.42), we have the self-consistent moment

$$m = [1 - (2h/I)^2]^{1/2} = [(I/2h)^2 - 1]^{1/2}/(I/2|h|). \tag{3.44}$$

This is plotted in the right-hand panel of Fig. 3.8 as a function of $I/2|h|$. Remembering that $h(R) \to 0$ as $R \to \infty$, we see that it shows the same square root distance-dependence as that displayed by the numerical self-consistent solution of the local spin density functional Schrödinger equation in Fig. 3.6. Thus, as the hydrogen molecule is pulled apart, it moves from the singlet state $S = 0$ at equilibrium to the isolated free atoms in doublet states with $S = \frac{1}{2}$.

The local density approximation (LDA) binding energy curve in Fig. 3.6, which accurately follows the exact curve around equilibrium, can be approximated by the sum of five terms, namely

$$U(R) = [\Phi_{ol}(R) + \Phi_{xc}(R) + \Phi_{es}(R)] - 2|h(R)| + U_\infty, \tag{3.45}$$

which are shown in the left-hand panel of Fig. 3.9. $U_\infty$ is a constant which accounts for the fact that the LDA binding energy curve dissociates incorrectly at infinity. The two dominant contributions are those expected from our earlier treatment of the covalent bond (cf eqn (3.19)): the overlap repulsion, $\Phi_{ol}(R)$ and the covalent bonding $2h(R)$ (where the prefactor of 2 accounts for both electrons being in the bonding state). Both these terms are quantum mechanical in origin, as too is the much weaker

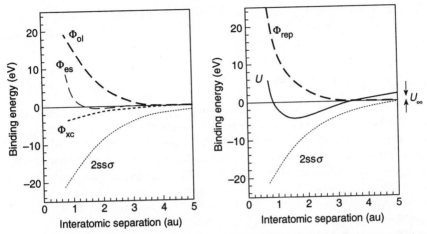

**Fig. 3.9** Left-hand panel: The overlap $\Phi_{ol}$, electrostatic $\Phi_{es}$, exchange-correlation $\Phi_{xc}$ and bond integral $2ss\sigma$ contributions to the binding energy of the hydrogen molecule (where $ss\sigma = h$). Right-hand panel: The binding energy curve (full line) is the sum of the three contributions $\Phi_{rep}$, $2ss\sigma$ and $U_\infty$ (see text for details). (After Skinner and Pettifor (1991).)

exchange-correlation contribution $\Phi_{xc}(R)$. The term $\Phi_{es}(R)$, on the other hand, is the classical electrostatic interaction between two hydrogen atoms in which the electronic charge densities are 'frozen' as the atoms come together to form the bond (cf the lower panel in Fig. 3.3). We see that this classical contribution imparts negligible binding at the equilibrium separation, although it does contribute noticeably to the repulsion at smaller distances when the two positive nuclei become less well screened from each other by the valence electrons.

We will simplify eqn (3.45) by representing the three terms in the square bracket by the single contribution $\Phi_{rep}(R)$. This is repulsive for $R < 5$ au, as shown in the right-hand panel of Fig. 3.9. We thus write

$$U(R) = \Phi_{rep}(R) - 2|h(R)| + U_\infty. \qquad (3.46)$$

Since the dominant contribution to $\Phi_{rep}$ is the overlap repulsion, we expect from eqn (3.19) that

$$\Phi_{rep}(R) \approx |h(R)|S(R) \approx A[h(R)]^2. \qquad (3.47)$$

The second equality follows by making the so-called Wolfsberg–Helmholtz approximation, that $S(R)$ is proportional to $h(R)$. Equation (3.47) is a good approximation for the hydrogen dimer around its equilibrium internuclear separation, since from Fig. 3.9 $\Phi_{rep} \alpha h^{2.1}$. We will see in the next chapter that the particular power-law relation between $\Phi_{rep}(R)$ and $h(R)$ plays an important role in determining the relative stability of different competing molecular structures types. We should note that a Lennard–Jones potential also has its repulsive $R^{-12}$ contribution varying as the square of its attractive $R^{-6}$ contribution, as in eqn (3.47).

## 3.5. $\sigma$, $\pi$, and $\delta$ bonds

A diatomic molecule has cylindrical symmetry about the internuclear axis, so that angular momentum is conserved in this direction. Quantum mechanically, this implies that the state of the molecule is characterized by the quantum number $m$ where $m\hbar$ gives the component of the angular momentum along the molecular axis. However, unlike the free atom, in which the $(2l + 1)$ different $m$ values are degenerate, the degeneracy is lifted in the molecule. By analogy, with the s, p, d, ... states of a free atom representing the orbital quantum numbers $l = 0, 1, 2, \ldots$, it is customary to refer to $\sigma$, $\pi$, $\delta$, ... states of a molecule as those corresponding to $m = 0, \pm1, \pm2, \ldots$ respectively.

Figure 3.10 illustrates the different characteristics of the $\sigma$, $\pi$, and $\delta$ bonds. We have seen from our previous discussion on the homonuclear diatomic molecule that a given atomic energy level will split into bonding and antibonding states separated by $2|h|$ where $h$ is the bond integral that couples $\psi_A$ and $\psi_B$ together through the average molecular potential $\bar{V}$ (cf. eqn (3.15)). If $\psi_A$ and $\psi_B$ are spherically symmetric s orbitals, then a ss$\sigma$ bond is formed as shown schematically in Fig. 3.10(a). If $\psi_A$ and $\psi_B$ are p orbitals, whose probability clouds are drawn in Fig. 2.15, then the three-fold degenerate free atom level (excluding spin degeneracy) splits into the singly degenerate pp$\sigma$ molecular state ($m = 0$) and the doubly degenerate pp$\pi$ molecular state ($m = \pm1$) as shown in Fig. 3.10(b). If $\psi_A$ and $\psi_B$ are d orbitals, whose probability clouds are sketched in Fig. 2.15, then the five-fold degenerate free atom level splits into the singly degenerate dd$\sigma$ molecular state ($m = 0$) and the two doubly degenerate molecular states dd$\pi$ ($m = \pm1$) and dd$\delta$ ($m = \pm2$) as shown in Fig. 3.10(c). For the case of a heteronuclear diatomic

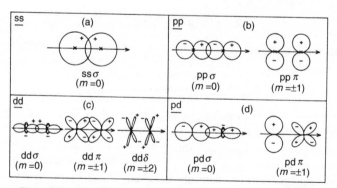

**Fig. 3.10** The formation of $\sigma$, $\pi$, and $\delta$ bonds (see text).

molecule, such as TiC, where the carbon p orbitals overlap the titanium d orbitals, a pd bond will be formed from the pdσ and pdπ states illustrated in Fig. 3.10(d).

It is clear from Fig. 3.10 that a σ bond is relatively strong, since the angular lobes point along the molecular axis and can give rise to a large overlap in the bonding region. On the other hand, the ppπ and ddδ bonds will be relatively much weaker since angular lobes extend in the plane perpendicular to the molecular axis. This can be seen explicitly by making the Wolfsberg–Helmholtz approximation and writing $h(R) = -S(R)$ in units of energy for each bond, such that the constant of proportionality is unity. Then for the overlap of hydrogenic-like 2s and 2p orbitals (cf eqs (2.51) and (2.52)), we have

$$ss\sigma = -(1 + \kappa R + \tfrac{4}{9}\kappa^2 R^2 + \tfrac{1}{9}\kappa^3 R^3 + \tfrac{1}{45}\kappa^4 R^4)\, e^{-\kappa R} \qquad (3.48)$$

$$sp\sigma = \tfrac{1}{6}\kappa R(1 + \kappa R + \kappa^2 R^2)\, e^{-\kappa R} \qquad (3.49)$$

$$pp\sigma = (-1 - \kappa R - \tfrac{1}{5}\kappa^2 R^2 + \tfrac{2}{15}\kappa^3 R^3 + \tfrac{1}{15}\kappa^4 R^4)\, e^{-\kappa R} \qquad (3.50)$$

$$pp\pi = -(1 + \kappa R - \tfrac{2}{5}\kappa^2 R^2 + \tfrac{1}{15}\kappa^3 R^3)\, e^{-\kappa R}. \qquad (3.51)$$

$\kappa^2$ is the magnitude of the appropriate atomic energy level and $R$ is the internuclear separation.

These are plotted in Fig. 3.11 and illustrate three important characteristics of s- and p-bond integrals. Firstly, ssσ and ppπ are negative due to

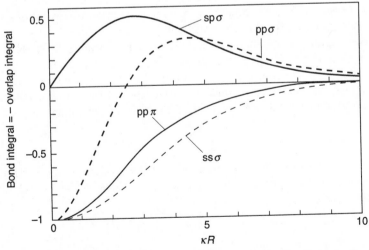

**Fig. 3.11** The bond integrals ssσ, spσ, ppσ, and ppπ as a function of κR where $\kappa^2$ is the magnitude of the appropriate valence energy level and $R$ is the internuclear separation (see eqs (3.48)–(3.51) and the text for details).

lobes of the same sign overlapping in the negative molecular potential, whereas spσ and ppσ (at intermediate and large distances) are positive due to lobes of opposite sign overlapping in the negative molecular potential (cf Fig. 3.10). Secondly, ppσ saturates as the bond distance decreases and changes sign at small internuclear separations because eventually there is more overlap between orbital regions of identical sign than of opposite (cf Fig. 3.10(b)). Thirdly, spσ also saturates as the bond distance decreases, since it must eventually vanish at $R = 0$ due to s and p orbitals on the same site being orthogonal. It follows from this behaviour that in the vicinity where ppσ saturates, as is observed for carbon and silicon bonds at their equilibrium separation, we have from Fig. 3.11 that

$$\text{ppσ} \approx \text{spσ} \approx |\text{ssσ}| \tag{3.52}$$

and

$$|\text{ppπ}| \approx \tfrac{1}{2}|\text{ssσ}|. \tag{3.53}$$

We will use these values when discussing the behaviour of sp-valent dimers in the next section.

## 3.6. Bond formation in sp-valent dimers

The binding properties across a given row of sp-valent dimers show the very marked triangular variation with valence shell occupancy that is observed in Fig. 3.12 for the 2s, 2p valent diatomic molecules. We see that the nitrogen molecule with its five valence electrons per atom displays the maximum binding energy, minimum internuclear separation, and maximum vibrational frequency (or curvature of the binding energy curve about equilibrium). We can understand this behaviour by extending our earlier treatment of s-valent diatomic molecules to that for sp-valent.

The cylindrical symmetry about the inter-nuclear axis leads to the solutions of the molecular Schrödinger equation, eqn (3.3), having either σ or π character. Taking the z axis along the axis of the molecule, the σ eigenfunctions will comprise linear combinations of the $\psi_{A_s}$, $\psi_{A_z}$, $\psi_{B_s}$, and $\psi_{B_z}$ atomic orbitals so that we can write the molecular orbital as

$$\psi_{AB} = \sum_{\alpha=s,z} (c_{A_\alpha}\psi_{A_\alpha} + c_{B_\alpha}\psi_{B_\alpha}) \tag{3.54}$$

By analogy with our previous treatment of the s-valent dimer, substituting eqn (3.54) into the Schrödinger equation (3.3) leads to the LCAO secular equation (neglecting the overlap integral, $S$)

$$\begin{bmatrix} E_s - E & 0 & \text{ssσ} & \text{spσ} \\ 0 & E_p - E & -\text{spσ} & \text{ppσ} \\ \text{ssσ} & -\text{spσ} & E_s - E & 0 \\ \text{spσ} & \text{ppσ} & 0 & E_p - E \end{bmatrix} \begin{bmatrix} c_{A_s} \\ c_{A_z} \\ c_{B_s} \\ c_{B_z} \end{bmatrix} = 0. \tag{3.55}$$

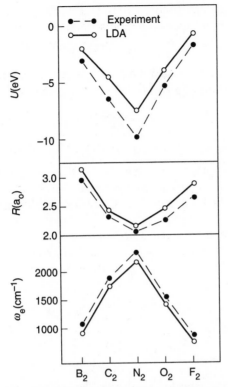

**Fig. 3.12** The binding energies, equilibrium internuclear separations and vibrational frequencies across the first-row diatomic molecules. Note the good agreement between the self-consistent local density approximation calculations and experiment for $R$ and $\omega_e$ but the larger systematic error of up to 2 eV for the binding energy. (After Gunnarsson *et al.* (1977).)

Since the potential is unchanged with respect to inversion about the centre of the molecule (i.e. $V_{AB}(\mathbf{r}) = V_{AB}(-\mathbf{r})$, with the origin at the molecular centre), the solutions will be either even (*gerade*) or odd (*ungerade*). The even solutions correspond to $(c_{A_s} = c_{B_s}, c_{A_z} = -c_{B_z})$, the odd solutions to $(c_{A_s} = -c_{B_s}, c_{A_z} = c_{B_z})$. Substituting into the LCAO secular equation above, we find that the even solutions are given by

$$\begin{bmatrix} E_s + ss\sigma - E & -sp\sigma \\ -sp\sigma & E_p - pp\sigma - E \end{bmatrix} \begin{bmatrix} c_{A_s} \\ c_{A_z} \end{bmatrix} = 0 \qquad (3.56)$$

whereas the odd solutions are given by

$$\begin{bmatrix} E_s - ss\sigma - E & sp\sigma \\ sp\sigma & E_p + pp\sigma - E \end{bmatrix} \begin{bmatrix} c_{A_s} \\ c_{A_z} \end{bmatrix} = 0. \qquad (3.57)$$

The resultant two quadratic equations can be solved directly. However, to keep the expressions simple without affecting our fundamental under-standing, we make the approximation suggested by eqn (3.52), namely

$$\text{pp}\sigma = \text{sp}\sigma = -\text{ss}\sigma = -h, \tag{3.58}$$

so that $h < 0$ as for the s-valent dimer. The even (g) and odd (u) eigenvalues, resulting from substituting eqn (3.58) into eqs (3.56) and (3.57), are

$$E_g = \bar{E} - |h| \pm \tfrac{1}{2}[4h^2 + (\Delta E_{sp})^2]^{1/2} \tag{3.59}$$

and

$$E_u = \bar{E} + |h| \pm \tfrac{1}{2}[4h^2 + (\Delta E_{sp})^2]^{1/2}, \tag{3.60}$$

where $\bar{E} = \tfrac{1}{2}(E_s + E_p)$ is the average value of the s and p atomic energy levels and $\Delta E_{sp} = E_p - E_s$ is the *splitting* between the free atom s and p valence energy levels (given by comparing the curves on the right- and left-hand sides of Fig. 2.16).

The behaviour of these eigenvalues can best be understood by considering the two limiting cases: (i) $\Delta E_{sp}/|h| \to \infty$ and (ii) $\Delta E_{sp}/|h| \to 0$. For the case where the sp splitting is large we have from eqs (3.59) and (3.60) that

$$E_g \to \begin{cases} E_s - |h| \\ E_p - |h| \end{cases}$$

$$\text{as } \Delta E_{sp}/|h| \to \infty \tag{3.61}$$

$$E_u \to \begin{cases} E_s + |h| \\ E_p + |h|. \end{cases}$$

Thus, we recover the expected result that the free atomic s and p levels split independently of each other into bonding and antibonding states, with energies $E_s \mp \text{ss}\sigma$ and $E_p \pm \text{pp}\sigma$ respectively (since $\text{ss}\sigma = h$, $\text{pp}\sigma = -h$). For the case where the sp splitting is small, on the other hand, we have

$$E_g \to \begin{cases} \bar{E} - 2|h| \\ \bar{E} \end{cases}$$

$$\text{as } \Delta E_{sp}/|h| \to 0 \tag{3.62}$$

$$E_u \to \begin{cases} \bar{E} \\ \bar{E} + 2|h|. \end{cases}$$

Thus, in this case where the s and p energy levels mix together or hybridize, we recover the bonding state $\bar{E} - 2|h|$, the antibonding state $\bar{E} + 2|h|$, and the doubly degenerate non-bonding state, $\bar{E}$. (Note that for the particular case where $\text{sp}\sigma$ is given by the geometric mean of the magnitudes of $\text{ss}\sigma$ and $\text{pp}\sigma$, that is $\text{sp}\sigma = (|\text{ss}\sigma|\text{pp}\sigma)^{1/2}$, the bonding and antibonding states for $\Delta E_{sp} = 0$ are given by $\bar{E} \pm (|\text{ss}\sigma| + \text{pp}\sigma)$, which for $|\text{ss}\sigma| = \text{pp}\sigma = |h|$ is consistent with eqn (3.62) above).

This behaviour is illustrated schematically by the left-hand panel of Fig. 3.13. We see that for large internuclear separations the $\sigma$ states go over to bonding and antibonding states around the free atomic energy levels $E_s$

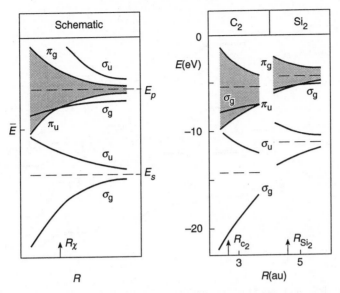

**Fig. 3.13** Left-hand panel: The electronic structure of an sp-valent diatomic molecule as a function of the internuclear separation. Labels $E_s$ and $E_p$ mark the positions of the free-atomic valence s and p levels respectively and $\bar{E} = \frac{1}{2}(E_s + E_p)$. The quantity $R_\chi$ is the distance at which the $\pi_u$ and upper $\sigma_g$ levels cross. The region between the upper and lower $\pi$ levels has been shaded to emphasize the increase in their separation with decreasing distance that is responsible for this crossing. Right-hand panel: The self-consistent local density approximation electronic structure for $C_2$ and $Si_2$ whose equilibrium internuclear separations are marked by $R_{C_2}$ and $R_{Si_2}$ respectively. (After Harris 1984.)

and $E_p$ respectively. However, as the two atoms come together, the bond integral increases, causing $\Delta E_{sp}/|h|$ to decrease and the s and p states to hybridize. As a result, as the internuclear separation decreases, the initially antibonding s state $\sigma_u$ and bonding p state $\sigma_g$ tend towards the nonbonding state $\bar{E}$. We have also sketched the doubly degenerate $\pi$ levels that correspond to the overlap of the $p_x$ and $p_y$ orbitals on the two atoms. We observe that at large separations they split about the atomic p level, $E_p$, with a width that is approximately half that of the $\sigma$ states as suggested by eqn (3.53). Note that because of the symmetry of the $\pi$ orbitals, which are displayed in Fig. 3.10(b), the bonding state is odd under inversion about the molecular centre, the antibonding state even. Hence, they are labelled $\pi_u$ and $\pi_g$ respectively.

A very important feature of the energy levels in Fig. 3.13 is the crossing of the $\pi_u$ and upper $\sigma_g$ levels at the internuclear separation $R_\chi$. This crossing must occur as the bond length gets smaller since the upper $\sigma_g$ level is bounded from below by $\bar{E}$ (cf eqn (3.62)), whereas the $\pi_u$ level carries on falling due

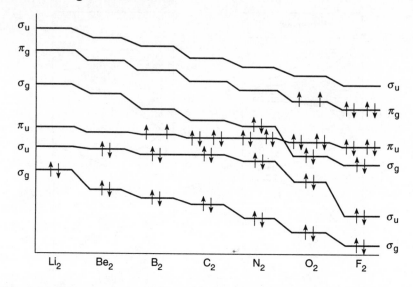

**Fig. 3.14** The occupancy of the molecular orbital energy levels across the first row diatomic molecules. (After Cotton and Wilkinson (1980).)

to $|pp\pi|$ increasing with decreasing distance (cf Fig. 3.11). As illustrated schematically by Fig. 3.14, the $\pi_u$ level lies below the upper $\sigma_g$ level for the sp-valent dimers to the left of nitrogen, whereas it lies above for the sp-valent dimers to the right of nitrogen. This switch in ordering of the $\pi_u$ and upper $\sigma_g$ levels as one moves across the period is not unexpected, since $\Delta E_{sp}/|h|$ increases on going across from $Li_2$ to $F_2$ (cf Fig. 2.16).

The ordering of the energy levels in Fig. 3.14 allows us to understand the cohesive properties of the sp-valent dimers that are displayed in Fig. 3.12. The two valence electrons in $Li_2$ both go into the lowest bonding state $\sigma_g$, whereas the extra two valence electrons in $Be_2$ would occupy the antibonding state $\sigma_u$, thereby leading to zero net bonding. Thus, although $Li_2$ is a stable molecule with about 1 eV of binding energy, $Be_2$ is not observed in the gaseous state. The additional two electrons in $B_2$ occupy the doubly degenerate $\pi_u$ state with both spins parallel by Hund's rule. Hence, whereas $Li_2$ is a diamagnetic molecule, $B_2$ is observed to be paramagnetic. $B_2$ will have a bond strength of unity, since it has a pair of electrons in the bonding $\pi$ state, with the four electrons in the bonding and antibonding $\sigma$ states contributing zero. The doubly degenerate $\pi_u$ state is fully occupied by four electrons in $C_2$ so that the carbon dimer is diamagnetic with a bond strength of two. The further two electrons in $N_2$ occupy the upper $\sigma_g$ level, thereby contributing a further unit to the bond strength (cf Fig. 3.13). Thus, the nitrogen dimer has a total bond strength of three, which leads it to having the largest binding energy and smallest equilibrium internuclear separation of the sp-valent series in Fig. 3.12.

The ordering of the $\pi_u$ and upper $\sigma_g$ levels switches over at oxygen. The extra two electrons in $O_2$ go into the doubly degenerate antibonding $\pi_g$ level with parallel spins, following Hund's rule, so that the oxygen dimer is paramagnetic as observed with a bond strength of two. Finally, the additional two electrons in $F_2$ fully occupy the antibonding $\pi_g$ level, causing the fluorine dimer to be diamagnetic with a bond strength of only unity. The variation in bond strength from $1 \rightarrow 2 \rightarrow 3 \rightarrow 2 \rightarrow 1$ as we go across from $B_2 \rightarrow C_2 \rightarrow N_2 \rightarrow O_2 \rightarrow F_2$ accounts beautifully for the observed variation in cohesion and bond length *across* the sp-valent series in Fig. 3.12.

The right-hand panel of Fig. 3.13 shows the variation of the electronic structure as we go *down* the group IV column from $C_2$ to $Si_2$. We see that if we were to shift the valence s and p atomic energy levels for silicon so that the corresponding horizontal dashed lines joined smoothly onto those of carbon, then the electronic structure would vary fairly smoothly from silicon to carbon, the bonding–antibonding energy level separations increasing with decreasing distance. The behaviour of their electronic structure, therefore, is not markedly different, $C_2$ having larger bond integrals than $Si_2$ due to its shorter bond length.

The crucial difference between $C_2$ and $Si_2$ is that at their equilibrium inter-nuclear separations, the relative ordering of the $\pi_u$ and upper $\sigma_g$ levels are reversed. This is seen from the fact that in Fig. 3.13

$$R_{C_2} < R_\chi \leq R_{Si_2}. \tag{3.63}$$

We should note in passing that the electronic structure is sensitive to the particular electronic configuration chosen through the self-consistent potential $V_{AB}$ in eqn (3.3). The right-hand panel of Fig. 3.13 has been drawn for the $\pi_u$ and upper $\sigma_g$ levels each being doubly occupied as $\pi_u^2\sigma_g^2$. If instead, all four electrons had been placed in the $\pi_u$ state, as for $C_2$ in Fig. 3.14, then this $\pi_u^4$ configuration would have caused the $\pi_u$ level to shift upwards and the $\sigma_g$ state to move downwards compared to their positions in Fig. 3.13, so that $R_\chi(\pi_u^4) < R_\chi(\pi_u^2\sigma_g^2)$.

The relative ordering that is observed in eqn (3.63) is due to the anomalously small core size of the carbon atom (cf Fig. 2.13). This allows the carbon atoms to approach together much more closely than might be expected from a simple extrapolation of the bond lengths of the other isovalent elements in group IV. Since for the carbon dimer at equilibrium the $\pi_u$ level lies below the $\sigma_g$ level, it is found that placing all four electrons in the pi state is more stable than placing two electrons in the pi state and two in the sigma state. On the other hand, for the silicon dimer at equilibrium the pure pi configuration $\pi_u^4$ is less stable by 1.5 eV than the mixed sigma-pi configuration $\pi_u^2\sigma_g^2$. Thus, the well-known fact that carbon favours pi bonded configurations whereas silicon prefers sigma bonded configurations may be traced to the anomalously small core size of carbon leading to a switching in the relative ordering of the $\sigma_g$ and $\pi_u$ energy levels at the equilibrium bond length.

## 3.7. Hybrid orbitals

The s and $p_z$ orbitals mix or hybridize as the free atoms are brought together to form the molecular bond. This can be seen directly from the behaviour of the $\sigma_g$ eigenfunction that corresponds to the most bonding state in Fig. 3.13. Substituting eqn (3.59) into eqn (3.56) we find

$$c_{A_s} = c_{B_s} = \frac{1}{\sqrt{2}} [1 + \Delta_{sp}/(1 + \Delta_{sp}^2)^{1/2}]^{1/2} \qquad (3.64)$$

and

$$c_{A_z} = -c_{B_z} = \frac{1}{\sqrt{2}} [1 - \Delta_{sp}/(1 + \Delta_{sp}^2)^{1/2}]^{1/2}, \qquad (3.65)$$

where $\Delta_{sp} = \Delta E_{sp}/2|h|$. Thus, from eqn (3.54) the most bonding eigenfunction can be written in the form

$$\psi_{AB} = \frac{1}{\sqrt{2}} (\phi_A + \phi_B), \qquad (3.66)$$

where $\phi_A$ and $\phi_B$ are the hybrid orbitals

$$\phi_A = \frac{1}{\sqrt{2}} [1 + \Delta_{sp}/(1 + \Delta_{sp}^2)^{1/2}]^{1/2} \psi_{A_s} + \frac{1}{\sqrt{2}} [1 - \Delta_{sp}/(1 + \Delta_{sp}^2)^{1/2}]^{1/2} \psi_{A_z}$$

$$(3.67)$$

$$\phi_B = \frac{1}{\sqrt{2}} [1 + \Delta_{sp}/(1 + \Delta_{sp}^2)^{1/2}]^{1/2} \psi_{B_s} - \frac{1}{\sqrt{2}} [1 - \Delta_{sp}/(1 + \Delta_{sp}^2)^{1/2}]^{1/2} \psi_{B_z}.$$

$$(3.68)$$

These are shown in Fig. 3.15 for the case of maximum mixing or hybridization that occurs for $\Delta_{sp} = 0$, when $E_s = E_p$, so that the two hybrids are given by $(1/\sqrt{2})(\psi_s \pm \psi_z)$. It follows from eqs (3.67) and (3.68) that the fraction of s character in the hybrid is given by $\frac{1}{2}[1 + \Delta_{sp}/(1 + \Delta_{sp}^2)^{1/2}]$ which, therefore, varies from 0.5 for $\Delta_{sp} = 0$ to 1 for $\Delta_{sp} = \infty$.

We see from eqn (3.66) that the bonding combination of the two sp hybrids $(1/\sqrt{2})(\psi_s + \psi_z)$ and $(1/\sqrt{2})(\psi_s - \psi_z)$ does *not* give the correct solution in general due to the non-vanishing sp splitting. (This splitting can be large, cf $C_2$ and $Si_2$ in Fig. 3.13.) Although these are the most directed hybrids that lead to the largest overlap and bond integral (for the particular choice $|ss\sigma| = pp\sigma$), a penalty has to be paid because electrons must be *promoted* from the low-lying s state into the high-lying p state that is mixed in. The

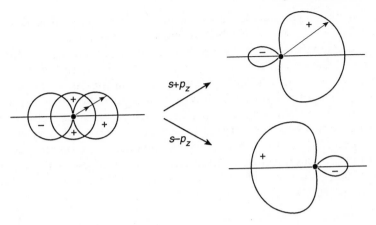

**Fig. 3.15** Formation of the two sp hybrids, $(1/\sqrt{2})(\psi_s \pm \psi_z)$. (After Huheey *et al.* (1993).)

eigenfunction optimizes the energy gain from overlap against the energy loss from promotion.

In fact, we can split the energy of the most bonding state into two parts as

$$E_{\sigma_g} = \frac{1}{2}\int (\phi_A + \phi_B)\hat{H}(\phi_A + \phi_B)\,d\mathbf{r} = \tfrac{1}{2}(H_{AA} + H_{BB}) + H_{AB} \quad (3.69)$$

where $H_{AA}$, $H_{BB}$, and $H_{AB}$ are the Hamiltonian matrix elements with respect to the hybrid orbitals $\phi_A$ and $\phi_B$. Substituting eqs (3.67) and (3.68) into (3.66) and neglecting the overlap contributions we find

$$\tfrac{1}{2}(H_{AA} + H_{BB}) = c_{A_s}^2 E_s + c_{A_z}^2 E_p = \bar{E} - \frac{\Delta_{sp}^2}{(1 + \Delta_{sp}^2)^{1/2}}|h| \quad (3.70)$$

and

$$H_{AB} = -\left[1 + \frac{1}{(1 + \Delta_{sp}^2)^{1/2}}\right]|h|, \quad (3.71)$$

where we have used eqn (3.58), namely $pp\sigma = sp\sigma = -ss\sigma = -h$. Thus, adding eqs (3.70) and (3.71) we recover

$$E_{\sigma_g} = \bar{E} - |h| - (1 + \Delta_{sp}^2)^{1/2}|h| \quad (3.72)$$

as expected from eqn (3.59). This energy is lower than the energy of $\bar{E} - 2|h|$ that would have resulted from using the sp hybrids, $(1/\sqrt{2})(\psi_s \pm \psi_z)$. Although the bond integral $|H_{AB}|$ is decreased from $2|h|$ to

$$[1 + 1/(1 + \Delta_{sp}^2)^{1/2}]|h|,$$

the cost in promoting electrons into p states is reduced by the larger amount

$[\Delta_{sp}^2/(1 + \Delta_{sp}^2)^{1/2}]|h|$. This leads to the total energy being lowered by the amount $[(1 + \Delta_{sp}^2)^{1/2} - 1]|h|$ which is observed in eqn (3.72).

It is commonly asserted that hybrid orbitals are responsible for the ground-state structures that are observed. For example, carbon takes the graphitic structure due to $sp^2$ hybrids, whereas silicon takes the diamond structure due to $sp^3$ hybrids. However, as has been shown by the above discussion, the actual hybrid chosen is a delicate balance between two opposing factors—the one trying to maximize the bond overlap with neighbouring atoms, the other trying to minimize the promotion energy. Structural prediction requires a careful treatment of the different competing terms in the total energy. We address this area for the case of molecules in the next chapter.

# References

Cotton, F. A. and Wilkinson, G. (1980). *Basic inorganic chemistry.* Wiley, New York.

Gunnarsson, O., Harris, J., and Jones, R. O. (1977). *The Journal of Chemical Physics* **67**, 3970.

Gunnarsson, O. and Lundquist, B. I. (1976). *Physical Review* **B13**, 4274.

Harris, J. (1984). *The electronic structure of complex systems* (ed. P. Phariseau and W. M. Temmerman), page 141. Plenum Press, New York.

Huheey, J. E., Keiter, E. A., and Keiter, R. L. (1993). *Inorganic chemistry: principles of structure and reactivity.* Harper Collins, New York.

Mulliken, R. S. (1934). *Journal of Chemical Physics* **2**, 782.

Pauling, L. (1960). *The nature of the chemical bond.* Cornell University, Ithaca, New York State.

Phillips, J. C. and Van Vechten, J. A. (1969). *Physical Review Letters* **22**, 705.

Skinner, A. J. and Pettifor, D. G. (1991). *Journal of Physics: Condensed Matter* **3**, 2029.

# 4
# Structure of molecules

## 4.1. Introduction

Molecules show a wide variety of structure, the number of variants increasing dramatically with number of atoms in the molecule. Thus, whereas the dimer displays only one structure type, the trimer can be either linear like $CO_2$ or bent like $H_2O$. Molecules with four atoms, on the other hand, can take the one-dimensional linear chain structures like $H_4$, the two-dimensional square ring structure like $S_4$ or the rhombic close-packed planar structure like $Si_4$, the three-dimensional tetrahedral structure like $P_4$, or numerous other distorted structural variants. The five-atom molecule, as we have already seen in Fig. 1.15, chooses the close-packed plane as its ground state structure for $Na_5$, the trigonal bipyramid for $Mg_5$ and $Si_5$, the regular pyramid for $Al_5$, and the envelope structure for $S_5$ respectively. The six-atom molecule can choose between the one-dimensional linear chain, the two-dimensional hexagonal ring or close-packed plane, the three-dimensional pentagonal pyramid, trigonal tripyramid, trigonal prism, or octahedron—to name but a few!

The energy difference between these competing structure types is usually very small, often being of the order of 1% of the binding energy or less. Hence, the reliable prediction of the ground-state structure of a *particular* molecule requires self-consistent quantum mechanical calculations of high precision that expend large amounts of computer time. In this chapter we will instead extend the simple ideas of dimeric bonding to larger molecules with up to six atoms, in order to study the structural *trends* that are observed as a function of the electron count or total number of valence electrons. We will see that the oscillatory trends in structural stability can be linked directly to the topology of the molecule through a very important moments theorem, which was first introduced into this area by Ducastelle and Cyrot-Lackmann in 1971.

## 4.2. Structural stability: an illustrative example

The relative stability of different structures is determined by a delicate balance between competing contributions to the total energy. This is illustrated by the following example in which we examine the stability of four-atom molecules with respect to the ideal linear chain, square, rhombus

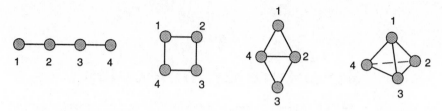

**Fig. 4.1** The linear chain, square, rhombus, and tetrahedron geometries of four-atom molecules.

or tetrahedron as shown in Fig. 4.1. We will simplify the problem by assuming that the energetics is describable by inter-atomic pair potentials so that the binding energy of the molecule may be written

$$U = \tfrac{1}{2} {\sum_{i,j}}' \Phi(R_{ij}), \qquad (4.1)$$

where $\Phi(R_{ij})$ is a central pair potential that is only dependent on the internuclear separation, $R_{ij}$, between atoms i and j in the molecule.

Following our discussion of the dimeric bond centred on eqn (3.46), we will divide up the pair potential into a repulsive and a bonding contribution, namely

$$\Phi(R) = \Phi_{\text{rep}}(R) + \Phi_{\text{bond}}(R). \qquad (4.2)$$

(The constant shift, $U_{\infty}$, in eqn (3.46) need not concern us here since it is the *relative* stability of the different structures that we wish to predict in this chapter.) Around the equilibrium bond length, $R_0$, the two contributions in eqn (4.2) may be related by

$$\Phi_{\text{rep}}(R) = A|\Phi_{\text{bond}}(R)|^{\lambda}, \qquad (4.3)$$

where $A$ is a constant and the exponent $\lambda$ is given explicitly by

$$\lambda = \{(\Phi'_{\text{rep}}/\Phi_{\text{rep}})/(\Phi'_{\text{bond}}/\Phi_{\text{bond}})\}_{R=R_0}. \qquad (4.4)$$

This can be expressed as the ratio of dimensionless logarithmic derivatives, namely

$$\lambda = \left\{ \left[ R \frac{\mathrm{d}}{\mathrm{d}R} (\ln \Phi_{\text{rep}}) \right] \middle/ \left[ R \frac{\mathrm{d}}{\mathrm{d}R} (\ln \Phi_{\text{bond}}) \right] \right\}_{R=R_0}. \qquad (4.5)$$

It follows from eqs (4.2) and (4.3) that the curvature of the potential about equilibrium can be written as

$$\Phi''(R_0) = (\lambda - 1)n^2|\Phi_{\text{bond}}(R_0)|/R_0^2, \qquad (4.6)$$

where $n$ is the dimensionless logarithmic derivative of the bonding contribution to the potential. Hence, we have a normalized curvature about the equilibrium bond length $R_0$ that is given by

$$\hat{\Phi}''(R_0) = [R_0^2/n^2|\Phi_{\text{bond}}(R_0)|]\Phi''(R_0) = (\lambda - 1). \qquad (4.7)$$

We will define the *degree of normalized hardness* of an interatomic potential by

$$\alpha_h = \frac{(\lambda - 1)}{\lambda}. \qquad (4.8)$$

This normalized hardness scale runs from zero (for a totally soft potential when $\lambda = 1$) to unity (for a totally hard potential when $\lambda = \infty$).

For evaluating the binding energy curves of the four-atom molecules, we now assume that the bonding potential falls off algebraically with distance, in particular

$$\Phi_{\text{bond}}(R) = -B/R^2. \qquad (4.9)$$

This inverse-square dependence is chosen because the magnitude of the $ss\sigma$ bond integral in Fig. 3.11 decreases approximately as $R^{-2}$ in the vicinity of $\kappa R \approx 5$. We are identifying the attractive contribution to the pair potential with covalent bonding as in eqn (3.46). From eqn (4.2) we have

$$\Phi(R) = AB^{\lambda}/R^{2\lambda} - B/R^2, \qquad (4.10)$$

or in terms of the more physically transparent parameters, $\varepsilon$ and $R_h$,

$$\Phi(R) = \varepsilon\left[\left(\frac{R_h}{R}\right)^{2\lambda} - \left(\frac{R_h}{R}\right)^2\right], \qquad (4.11)$$

where $\varepsilon = A^{-1/(\lambda-1)}$ and $R_h = A^{1/2(\lambda-1)}B^{1/2}$. Thus, $\varepsilon$ sets the energy scale, whereas $R_h$ sets the length scale. The pair potential, $\Phi/\varepsilon$, is plotted in Fig. 4.2 as a function of $R/R_h$ for different values of $\lambda$ or degree of normalized hardness, $\alpha_h$. We see that for $\alpha_h = 0$ corresponding to $\lambda = 1$, the potential is totally soft (in fact $\Phi(R)$ vanishes everywhere!) whereas for $\alpha_h = 1$ corresponding to $\lambda = \infty$ the potential is totally hard (in fact $\Phi(R)$ is infinite inside $R_h$, corresponding to twice the hard core radius).

The binding energy curves for the four-atom molecules shown in Fig. 4.1 will be sensitive to the degree of normalized hardness, $\alpha_h$. Summing over all the bonds in eqn (4.1), the total binding energies of the tetrahedron (t), rhombus (r), square (s), and linear chain (l) are given by

$$U^t(R) = 6\Phi(R), \qquad (4.12)$$

$$U^r(R) = 5\Phi(R) + \Phi(\sqrt{3}R), \qquad (4.13)$$

$$U^s(R) = 4\Phi(R) + 2\Phi(\sqrt{2}R), \qquad (4.14)$$

and

$$U^l(R) = 3\Phi(R) + 2\Phi(2R) + \Phi(3R). \qquad (4.15)$$

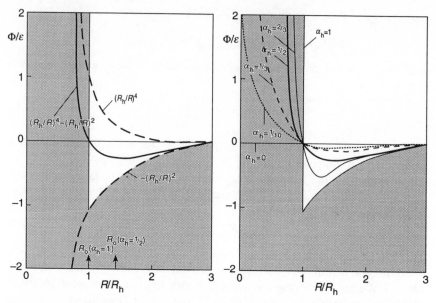

**Fig. 4.2** (a) The full curve shows the normalized pair potential, $\Phi/\varepsilon$, versus the normalized interatomic distance, $R/R_h$, for the degree of normalized hardness, $\alpha_h = \frac{1}{2}$, corresponding to $\lambda = 2$. The two dashed curves show the repulsive and attractive contributions respectively. The shaded region delineates the hard-core potential with $\alpha_h = 1$ corresponding to $\lambda = \infty$. The two vertical arrows mark the equilibrium nearest-neighbour distances for $\alpha_h = 1$ and $\frac{1}{2}$ respectively. (b) The normalized pair potential, $\Phi/\varepsilon$, versus the normalized interatomic distance for different values of the degree of normalized hardness, $\alpha_h$. Note that $\alpha_h = 0$ corresponds to a totally soft potential, $\alpha_h = 1$ to a totally hard potential.

Substituting eqn (4.11) into the above equations we can write them in the form

$$U^\gamma(R) = \varepsilon \left\{ [\varkappa_{rep}^\gamma(\lambda)] \left( \frac{R_h}{R} \right)^{2\lambda} - [\varkappa_{bond}^\gamma] \left( \frac{R_h}{R} \right)^2 \right\} \qquad (4.16)$$

with $\gamma = $ t, r, s, or l. $\varkappa_{rep}^\gamma$ and $\varkappa_{bond}^\gamma$ are effective coordination numbers for the repulsive and bonding contributions respectively. $\varkappa_{rep}^\gamma(\lambda)$ is dependent on $\lambda$, reflecting the $\lambda$ dependence of the repulsive potential. Table 4.1 gives the resultant values of $\varkappa_{bond}^\gamma$ and $\varkappa_{rep}^\gamma$, the latter for the degrees of normalized hardness, $\alpha_h = 0$, $\frac{1}{3}$, $\frac{1}{2}$, and 1, corresponding to $\lambda = 1$, 1.5, 2, and $\infty$ respectively.

Figure 4.3 shows that the relative energy differences between the tetrahedron, rhombus, square and linear chain are dependent on the degree of normalized hardness $\alpha_h$. For a hard-core potential with $\alpha_h = 1$, all four molecules take the same equilibrium bond length $R_h$. We see that the most stable four-atom molecule is the tetrahedron with six nearest neighbour

**Table 4.1** The values of the effective coordination numbers $\varkappa^\gamma_{bond}$ and $\varkappa^\gamma_{rep}$ for the tetrahedron (t), rhombus (r), square (s), and linear chain (l). The degrees of normalized hardness $\alpha_h = 0, \frac{1}{3}, \frac{1}{2}$, and 1 correspond to $\lambda = 1, 1.5, 2$, and $\infty$ respectively.

| $\gamma$ | $\varkappa^\gamma_{bond}$ | $\varkappa^\gamma_{rep}$ | | | |
|---|---|---|---|---|---|
| | | $\alpha_h = 0$ | $\alpha_h = \frac{1}{3}$ | $\alpha_h = \frac{1}{2}$ | $\alpha_h = 1$ |
| t | 6.000 | 6.000 | 6.000 | 6.000 | 6.000 |
| r | 5.333 | 5.333 | 5.192 | 5.111 | 5.000 |
| s | 5.000 | 5.000 | 4.707 | 4.500 | 4.000 |
| l | 3.611 | 3.611 | 3.287 | 3.137 | 3.000 |

bonds, followed in order by the rhombus with 5.333 effective bonds, the square with 5.000 effective bonds, and the linear chain with 3.611 effective bonds (cf Table 4.1). However, as the degree of normalized hardness of the potential decreases, we see that the more open structures take shorter bond lengths at equilibrium than the most close-packed tetrahedral atomic configuration. This gives increased cohesion to the more open structures. Although the tetrahedron always remains the most stable molecule for this pair potential description of bonding, the relative stability of the rhombus and the square switches over when the degree of normalized hardness changes from $\alpha_h = \frac{1}{2}$ to $\alpha_h = \frac{1}{3}$. Therefore, for a sufficiently soft interatomic potential, the more open planar structure, the square, becomes more stable than the close-packed planar structure, the rhombus.

As we have already mentioned in the previous chapter, the carbon core behaves anomalously. Although the carbon interatomic potential has a much stronger curvature about equilibrium than is the case for silicon, the bond length and inverse bond integrals that normalize the potential in eqn (4.7) are much smaller (cf Fig. 3.13). Hence we find that $\lambda$ takes a value of about 1.8 for carbon but about 2.2 for silicon, so that $\alpha^C_h/\alpha^{Si}_h = 0.444/0.545 = 0.81$. This relative softness of the normalized interatomic potentials of the first-row elements carbon, nitrogen, and oxygen will be shown, in the last chapter, to account for their taking more open ground-state structures than the other elements in their respective groups in the periodic table.

## 4.3. The structural energy difference theorem

The energy difference between two competing structure types is usually very small compared to the total binding energy. Moreover, the simplest assumption that the first nearest-neighbour bond length is structure-independent can lead to incorrect predictions. For example, for the degree

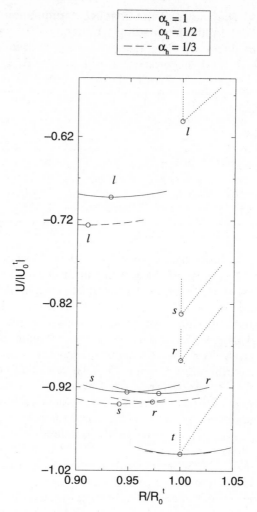

**Fig. 4.3** The normalized binding energy curves, $U/|U_0^t|$, versus the normalized nearest-neighbour bond length, $R/R_0^t$, for different values of the degree of normalized hardness, $\alpha_h$. Terms $U_0^t$ and $R_0^t$ are the equilibrium binding energy and nearest-neighbour bond length of the tetrahedron for a given value of $\alpha_h$.

of normalized hardness, $\alpha_h = \frac{1}{3}$, in Fig. 4.3, the square is found to be more or less stable than the rhombus, depending on whether the fixed bond length is taken as that of the equilibrium square or rhombus respectively. In order to separate out the delicate interplay between the repulsive and bonding contributions in establishing equilibrium bond lengths and hence energies, we now derive a theorem that expresses the structural energy differences within a well-defined two-step procedure (Pettifor (1986)).

In general, as we have seen, the binding energy can be written as the sum of a bonding and a repulsive contribution, namely

$$U = U_{\text{rep}} + U_{\text{bond}}. \tag{4.17}$$

The structural energy difference theorem states that the energy difference, $\Delta U$, between two structures is given to first order by

$$\Delta U^{(1)} = [\Delta U_{\text{bond}}]_{\Delta U_{\text{rep}} = 0}. \tag{4.18}$$

*Proof:* Consider two structures I and II with equilibrium first nearest-neighbour bond lengths, $R_0^{\text{I}}$ and $R_0^{\text{II}}$, respectively. Then

$$\Delta U = U^{\text{II}}(R_0^{\text{II}}) - U^{\text{I}}(R_0^{\text{I}}). \tag{4.19}$$

Let $R^{\text{II}}$ be the distance at which structure II displays the same repulsive energy as structure, I, at equilibrium, that is

$$U_{\text{rep}}^{\text{II}}(R^{\text{II}}) = U_{\text{rep}}^{\text{I}}(R_0^{\text{I}}). \tag{4.20}$$

But by expanding $U^{\text{II}}(R)$ in a Taylor series about $R_0^{\text{II}}$, we have

$$U^{\text{II}}(R^{\text{II}}) = U^{\text{II}}(R_0^{\text{II}}) + \frac{1}{2}\left(\frac{\mathrm{d}^2 U^{\text{II}}}{\mathrm{d}R^2}\right)_0 (R^{\text{II}} - R_0^{\text{II}})^2 + \cdots \tag{4.21}$$

where the linear term does not appear on the right-hand side since $\mathrm{d}U^{\text{II}}/\mathrm{d}R$ vanishes at the equilibrium separation $R_0^{\text{II}}$. Therefore, the substitution of eqn (4.21) into eqn (4.19) yields

$$\Delta U = U^{\text{II}}(R^{\text{II}}) - U^{\text{I}}(R_0^{\text{I}}) + O(R^{\text{II}} - R_0^{\text{II}})^2 \tag{4.22}$$

where the third term on the right-hand side represents second-order contributions. Consequently, substituting eqn (4.17) into eqn (4.22) and using the identity eqn (4.20), the energy difference is given to first order by

$$\Delta U^{(1)} = U_{\text{bond}}^{\text{II}}(R^{\text{II}}) - U_{\text{bond}}^{\text{I}}(R_0^{\text{I}}), \tag{4.23}$$

which may be expressed more compactly by eqn (4.18). This completes the proof.

This theorem is very important. It generalizes the usual procedure for studying the structural stability of ionic molecules or compounds by *first* packing together hard spheres until they touch and *then* comparing their electrostatic or Madelung energies in order to see which is most stable. The structural energy difference theorem allows us to extend this two-stage process to the case of realistic atoms or ions which do not exhibit hard-core behaviour (cf the left-hand panel of Fig. 4.2). In the first step, analogous to packing together hard spheres, the bond lengths of the competing structure types are adjusted to guarantee the same repulsive energy. In the second step, analogous to evaluating the ionic Madelung energies, the bond energies are computed and compared.

This theorem can be applied to our previous example of the relative stabilities of four-atom molecules. We shall compare the linear chain (l), square (s), and rhombus (r) with the tetrahedron (t) which we will take as our reference structure. From eqn (4.16) it has a bond length that is given by

$$R_0^t(\lambda) = \lambda^{1/2(\lambda-1)}R_h. \tag{4.24}$$

Thus, the tetrahedron has equilibrium bond lengths of $1.5R_h$, $\sqrt{2}R_h$, and $R_h$ for the degrees of normalized hardness $\alpha_h = \frac{1}{3}, \frac{1}{2}$, and 1 that correspond to $\lambda = 1.5, 2$, and $\infty$ respectively. As required by the first step of the theorem, we prepare the bond lengths, $R^\gamma$, of the three structures, $\gamma = $ l, s, and r, so that they display the same repulsive energies as the tetrahedron, t, that is

$$\varkappa_{rep}^\gamma/(R^\gamma)^{2\lambda} = \varkappa_{rep}^t/(R_0^t)^{2\lambda}. \tag{4.25}$$

Then, following the second step of the theorem, we compare the bond energies of the different structures, $\gamma$, at these prepared bond lengths with that of the tetrahedron, t, that is

$$\Delta U_{\gamma t}^{(1)} = -\varepsilon[\varkappa_{bond}^\gamma(R_h/R^\gamma)^2 - \varkappa_{bond}^t(R_h/R_0^t)^2]. \tag{4.26}$$

Substituting in for $R^\gamma$ from eqn (4.25) and for $R_0^t$ from eqn (4.24) and using $\varkappa_{bond}^t = \varkappa_{rep}^t = 6.000$ from Table 4.1, we have that the energy differences are given by

$$\Delta U_{\gamma t}^{(1)} = -\varepsilon\lambda^{1/(1-\lambda)}[(6/\varkappa_{rep}^\gamma)^{1/\lambda}\varkappa_{bond}^\gamma - 6], \tag{4.27}$$

where $\gamma = $ l, s, or r.

We see from Table 4.2 that the theorem gives the energy difference between the square or rhombus and the ground-state tetrahedron to within 2% of the exact values. Moreover, it predicts correctly the delicate switch in relative

**Table 4.2** The equilibrium energy difference $\Delta U = U_0^\gamma - U_0^t$, where $\gamma = $ r, s, or l. (t, r, s, and l correspond to the tetrahedron, rhombus, square and linear chain configurations respectively). The columns $\Delta U/\varepsilon$ give the exact energy differences, whereas the columns $\Delta U^{(1)}/\varepsilon$ give the first-order energy differences from the structural energy difference theorem eqn (4.18). The degrees of normalized hardness $\alpha_h = \frac{1}{3}, \frac{1}{2}$, and 1 correspond to $\lambda = 1.5, 2$, and $\infty$ respectively. The structural energy difference theorem is exact for $\alpha_h = 1$, so that $\Delta U^{(1)} = \Delta U$.

| $\gamma$ | $\alpha_h = \frac{1}{3}$ | | | $\alpha_h = \frac{1}{2}$ | | | $\alpha_h = 1$ | |
| | $\frac{\Delta U}{\varepsilon}$ | $\frac{\Delta U^{(1)}}{\varepsilon}$ | Error | $\frac{\Delta U}{\varepsilon}$ | $\frac{\Delta U^{(1)}}{\varepsilon}$ | Error | $\frac{\Delta U}{\varepsilon} =$ | $\frac{\Delta U^{(1)}}{\varepsilon}$ |
|---|---|---|---|---|---|---|---|---|
| r | 0.055 | 0.056(5) | 2% | 0.109 | 0.111 | 2% | 0.667 | |
| s | 0.053 | 0.054 | 2% | 0.111 | 0.113 | 2% | 1.000 | |
| l | 0.244 | 0.270 | 11% | 0.461 | 0.503 | 9% | 2.389 | |

stability between the rhombus and the square when the degree of normalized hardness changes from $\alpha_h = \frac{1}{2}$ to $\alpha_h = \frac{1}{3}$. This example demonstrates the accuracy of the structural energy difference theorem, which will be used extensively throughout the rest of this book.

## 4.4. The structure of s-valent molecules

An understanding of the structure of molecules requires a proper quantum mechanical description of the covalent bond that cannot be captured by the use of central pair potentials. We therefore extend our linear combination of atomic orbitals (LCAO) treatment of the s-valent dimer to three-, four-, five-, and six-atom molecules respectively. Following eqs (3.46) and (4.17), we write the binding energy *per atom* for an $\mathcal{N}$-atom molecule as

$$U = U_{rep} + U_{bond}. \tag{4.28}$$

The first term, $U_{rep}$, is an empirical pairwise repulsive contribution, namely

$$U_{rep} = \frac{1}{2\mathcal{N}} \sum_{i,j}' \Phi_{rep}(R_{ij}). \tag{4.29}$$

The second term $U_{bond}$ is the covalent bond energy that arises from occupying the molecular eigenstates, $n$, with electrons, namely

$$U_{bond} = \frac{1}{\mathcal{N}} \sum_n (E_n - E_s)f_n = \frac{1}{\mathcal{N}} \sum_n \varepsilon_n f_n, \tag{4.30}$$

where $f_n$ is the electron occupancy. The term $\varepsilon_n$ gives the energy of the eigenvalue $E_n$ with respect to the on-site atomic energy level $E_s$. Thus, $\varepsilon_n$ is negative for bonding states but positive for antibonding states. Equations (4.28)–(4.30) reduce to eqn (3.46) for the dimer when $\mathcal{N} = 2$, since in this case $\varepsilon_1 = h = ss\sigma$. (We will again ignore the constant shift in energy, $U_\infty$, in eqn (3.46) because we are interested in predicting the relative stability of different structures.)

The molecular eigenvalues are obtained by looking for the LCAO solution

$$\psi = \sum_{i=1,\mathcal{N}} c_i \psi_i \tag{4.31}$$

where $\psi_i$ is the atomic s orbital on site i. We shall evaluate the resultant LCAO secular equation within the so-called nearest-neighbour, orthogonal, two-centre tight binding (TB) approximation. This is indeed an appropriate connotation. Firstly, unlike our earlier illustrative example with pair potentials, we will assume that there is bonding only between *first* nearest neighbours within the molecule, since the valence orbitals are so tightly bound to their parent atoms. Secondly, we will ignore all overlap integrals, $S_{ij} = \int \psi_i \psi_j \, d\mathbf{r}$ (i $\neq$ j). Thirdly, we will neglect all three centre integrals of the form $\int \psi_i V_j \psi_k \, d\mathbf{r}$ (i $\neq$ j $\neq$ k).

The resultant TB eigenvalue equation takes the determinantal form,

$$|H - \varepsilon I| = 0, \tag{4.32}$$

where $H$ is the nearest-neighbour, two-centre Hamiltonian matrix and $I$ is the unit matrix. $H$ may be written down directly as we now illustrate for the four-atom molecules whose sites are labelled 1, 2, 3, and 4 as shown in Fig. 4.1. With nearest-neighbour bonding only, the linear chain, square, rhombus, and tetrahedron have

$$H^l = \begin{pmatrix} 0 & 1 & 0 & 0 \\ 1 & 0 & 1 & 0 \\ 0 & 1 & 0 & 1 \\ 0 & 0 & 1 & 0 \end{pmatrix} h^l, \tag{4.33}$$

$$H^s = \begin{pmatrix} 0 & 1 & 0 & 1 \\ 1 & 0 & 1 & 0 \\ 0 & 1 & 0 & 1 \\ 1 & 0 & 1 & 0 \end{pmatrix} h^s, \tag{4.34}$$

$$H^r = \begin{pmatrix} 0 & 1 & 0 & 1 \\ 1 & 0 & 1 & 1 \\ 0 & 1 & 0 & 1 \\ 1 & 1 & 1 & 0 \end{pmatrix} h^r, \tag{4.35}$$

and

$$H^t = \begin{pmatrix} 0 & 1 & 1 & 1 \\ 1 & 0 & 1 & 1 \\ 1 & 1 & 0 & 1 \\ 1 & 1 & 1 & 0 \end{pmatrix} h^t, \tag{4.36}$$

where $h^l$, $h^s$, $h^r$, and $h^t$ are the bond integrals for the respective molecules. Note that the diagonal elements of the above matrices vanish because the energy $\varepsilon$ is measured with respect to the on-site atomic energy level $E_s$.

The eigenvalue equation may, therefore, be solved to yield the four energy levels of each molecule, namely

$$\varepsilon^l = \pm \tfrac{1}{2}(1 \pm \sqrt{5})h^l, \tag{4.37}$$

$$\varepsilon^s = 0, 0, \pm 2h^s, \tag{4.38}$$

$$\varepsilon^r = 0, -h^r, \tfrac{1}{2}(1 \pm \sqrt{17})h^r, \tag{4.39}$$

and

$$\varepsilon^t = -h^t, -h^t, -h^t, 3h^t. \tag{4.40}$$

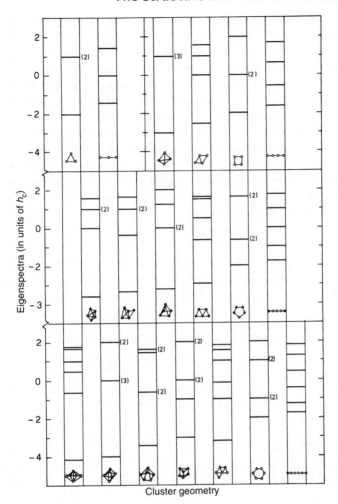

**Fig. 4.4** The eigenspectra of three-, four-, five-, and six-atom s-valent molecules in units of $|h^\gamma|$, where $h^\gamma$ is the nearest-neighbour bond integral for the given molecule $\gamma$. The numbers in parentheses give the degeneracy of the level. (From Shah and Pettifor (1993).)

These are plotted in the upper right-hand panel of Fig. 4.4, remembering that $ss\sigma$ integrals are negative. We see that the square has a pair of nonbonding states, whereas the tetrahedron has a triplet of antibonding states. We have also plotted the eigenspectra of three-atom, five-atom, and six-atom molecules. The molecules include all the most stable ground-state geometries of neutral and singly ionized s-valent systems, which were found by Wang *et al.* (1987) in their global search over *all* geometries constrained by equidistant nearest-neighbour bonds. Thus, we consider the linear chain and equilateral triangle for the three-atom molecules; the linear chain, square, rhombus, and tetrahedron for the four-atom molecules; the linear chain, pentagon, close-packed layer, square pyramid, monofinned tetrahedron, and trigonal bipyramid for the five-atom molecules; and the linear chain, hexagon, close-packed layer, trigonal prism, pentagonal pyramid, octahedron, and trigonal tripyramid for the six-atom molecules.

The relative stability of these molecules may be predicted using the structural energy difference theorem which we have proved in the previous section. That is, the differences in the total energies *per atom* are simply the differences in the bond energies provided that the bond lengths have first been adjusted, so that the molecules have identical repulsive energies. We will assume, as in eqn (4.3), that

$$\Phi_{rep}(R) = A|h(R)|^{\lambda}, \tag{4.41}$$

where $A$ is a constant. Then taking the dimer as the reference molecule with bond integral $h_0$, we have from eqn (4.29) on equating the repulsive energies per atom that

$$h^{\gamma} = [\mathcal{N}^{\gamma}/(2\mathcal{N}_b^{\gamma})]^{1/\lambda} h_0 \tag{4.42}$$

where $\mathcal{N}^{\gamma}$ and $\mathcal{N}_b^{\gamma}$ are the number of atoms and number of first nearest-neighbour bonds in cluster $\gamma$ respectively. Thus, for example, the $\mathcal{N} = 4$ linear chain, square, rhombus, and tetrahedron have $\mathcal{N}_b = 3, 4, 5$, and 6 respectively, whereas the $\mathcal{N} = 6$ trigonal prism and pentagonal pyramid have $\mathcal{N}_b = 9$ and 10 respectively. It follows from eqn (4.42) that in the hard-core limit with $\lambda = \infty$, the bond integrals take the same value for all molecules as expected, since the first nearest-neighbour bond length is invariant at twice the hard-core radius.

Figures 4.5 and 4.6 show the predicted bond energies per atom (in units of $|h_0|$) for three-, four-, five-, and six-atom clusters as a function of the electron count $N$ for the three different values of the degree of normalized hardness $\alpha_h = \frac{1}{2}, \frac{2}{3}$, and 1 (corresponding to $\lambda = 2, 3$, and $\infty$ respectively). The influence of the degree of normalized hardness, $\alpha_h$, is clearly illustrated by the upper panel of Fig. 4.5. We see that for $\alpha_h = 1$, when both three-atom molecules take the same nearest-neighbour bond length, and hence have identical bond integrals, $h_0$, the triangle is predicted to be the more stable molecule for the neutral monovalent system, whereas for $\alpha_h = \frac{1}{2}$ and $\frac{2}{3}$, the linear chain is more stable. Thus, whereas the alkalis with their relatively hard cores take a (Jahn–Teller distorted) triangular configuration, hydrogen remains linear, since $\alpha_h \approx \frac{1}{2}$ corresponding to $\lambda \approx 2$ (cf eqn (3.47)). As expected, decreasing the degree of normalized hardness favours less topologically close-packed structures with lower coordination.

The $\alpha_h = 1$ panels in Figs. 4.5 and 4.6 show that neutral three-, four-, five-, and six-atom molecules would take the triangle, rhombus, close-packed

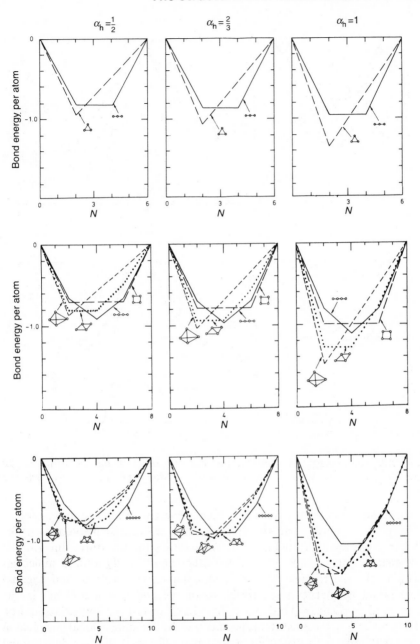

**Fig. 4.5** The average bond energy per atom (in units of the magnitude of the dimer bond integral $|h_0|$) as a function of the electron count $N$ for three-, four-, and five-atom molecules. The pentagon and square pyramid five-atom molecules have been omitted for clarity. (After Shah and Pettifor (1993).)

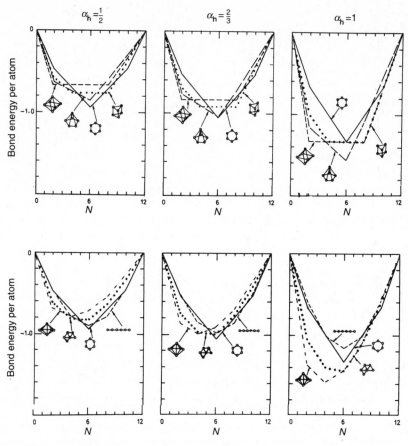

**Fig. 4.6** The average bond energy per atom (in units of $|h_0|$) as a function of the electron count, $N$, for six-atom molecules. (After Shah and Pettifor (1993).)

plane, and pentagonal pyramid as their respective ground-state structures. This is in agreement with the first-principles self-consistent predictions for Na that are recorded in Fig. 1.15. We see that the first three-dimensional geometry occurs for $\mathcal{N} = 6$. On the other hand, singly ionized molecules would take the triangle, rhombus, monofinned tetrahedron, and trigonal tripyramid geometries respectively. Again we find that a decrease in the degree of normalized hardness $\alpha_h$ stabilizes lower-coordinated structures. Thus, for $\alpha_h = \frac{1}{2}$, the most stable geometry is the one-dimensional linear chain for three, four-, and five-atom neutral molecules and the two-dimensional hexagonal ring for six-atom neutral molecules. First-principles calculations do indeed show, for example, that the symmetric linear chain for $H_4$ is about 1 eV more stable than the square geometry. We also see from Figs. 4.5 and 4.6 that for $\alpha_h = \frac{1}{2}$, the dimer has the lowest binding energy per atom, namely

$|h_0|$, whereas for $\alpha_h = 1$, the dimer has the highest energy of all the ground-state molecules. This is consistent with hydrogen being dimeric but the alkalis being close-packed metals.

## 4.5. Origin of structural trends: a moments theorem

What is the origin of the structural trends that are observed in Figs. 4.5 and 4.6 as a function of the number of valence electrons or electron count $N$? For example, why do four-atom molecules with $\alpha_h = \frac{1}{2}$ show the structural trend from tetrahedron → rhombus → linear chain → square as $N$ runs from 0 to 8? Moreover, why do the curves for the square and the linear chain cross each other twice whereas those for the tetrahedron and the linear chain cross only once? To answer these questions we now show that the moments of the eigenspectrum can be related directly to the topology of the molecule as was first done within a TB framework by Cyrot-Lackmann in 1967. We will then state a moments theorem that allows us to understand the origin of the structural trends in Figs. 4.5 and 4.6.

The $p$th moment of a given eigenspectrum $\{\varepsilon_n\}$ is defined by

$$\mu_p = \sum_n \varepsilon_n^p \qquad (4.43)$$

where $n$ runs over all $N$ states, whether occupied or unoccupied by electrons. But the Hamiltonian matrix, $H$, is diagonal with respect to the basis of eigenfunctions $\psi^{(n)}$, that is

$$H_{nm} = \varepsilon_n \delta_{nm}. \qquad (4.44)$$

Hence, we can write the $p$th moment as the trace over $H^p$, since

$$\mu_p = \sum_n (H_{nn})^p = \sum_n (H^p)_{nn} = \mathrm{Tr}\, H^p. \qquad (4.45)$$

However, as is well known, the trace is invariant with respect to choice of basis functions that are related by a unitary transformation. Thus, rather than working with the basis of eigenfunctions $\psi^{(n)}$, we may, following eqn (4.31), work with respect to the basis of atomic orbitals, $\psi_i$, to write

$$\mu_p = \mathrm{Tr}\, H^p = \sum_i (H^p)_{ii}. \qquad (4.46)$$

Performing the matrix multiplication explicitly we have

$$\mu_p = \sum_{i_1, i_2, \ldots, i_p} H_{i_1 i_2} H_{i_2 i_3} \cdots H_{i_p i_1}. \qquad (4.47)$$

We see, therefore, that the $p$th moment of the eigenspectrum is given by the sum over all bonding paths of length, $p$, that start and finish on the same atom within the molecule.

This is a key result since it links the moments of the eigenspectrum directly to the local topology of the molecule. Let us consider the first five moments, $\mu_0$, $\mu_1$, $\mu_2$, $\mu_3$, and $\mu_4$, respectively. The moment, $\mu_0$ gives *the total number of states* in the eigenspectrum or, equivalently, the total number of atomic orbitals, since from eqn (4.46)

$$\mu_0 = \text{Tr } H^0 = \text{Tr } I = \mathcal{N}. \tag{4.48}$$

The ratio $\mu_1/\mu_0$ gives the *centre of gravity* of the eigenspectrum. From eqn (4.47)

$$\mu_1/\mu_0 = \mathcal{N}^{-1} \sum_i H_{ii} = E_s = 0, \tag{4.49}$$

since $H_{ii} = E_s$ for all i, and we have taken $E_s$ as our present zero of energy (cf eqn (4.30)). The ratio $\mu_2/\mu_0$ gives the *mean square width* of the eigenspectrum. From eqn (4.47)

$$\mu_2/\mu_0 = \mathcal{N}^{-1} \sum_{ij} H_{ij} H_{ji} = \mathcal{N}^{-1} \sum_{i \neq j} h^2(R_{ij}), \tag{4.50}$$

as $H_{ii} = 0$ and $H_{ij} = h(R_{ij})$ for $i \neq j$ within the two-centre TB approximation. The ratio $\mu_3/\mu_0$ reflects the *skewness* of the eigenspectrum. From eqn (4.47)

$$\mu_3/\mu_0 = \mathcal{N}^{-1} \sum_{ijk} H_{ij} H_{jk} H_{ki} = \mathcal{N}^{-1} \sum_{i \neq j \neq k} h(R_{ij}) h(R_{jk}) h(R_{ki}). \tag{4.51}$$

The degree of skewness is measured by the dimensionless third moment, $\hat{\mu}_3/\hat{\mu}_2^{3/2}$, where the normalized moment $\hat{\mu}_p = \mu_p/\mu_0$ so that $\hat{\mu}_0 = 1$. The ratio $\mu_4/\mu_0$ gives a measure of *unimodal versus bimodal* behaviour in the eigenspectrum, as we will show later.

The four-atom molecules in Fig. 4.1 provide an excellent example of this direct connection between the moments of the eigenspectrum and the local topology. Firstly, as predicted by eqn (4.49), we see from Fig. 4.4 that the centre of gravity of all the eigenspectra is identically zero. Secondly, the root mean square width of the eigenspectrum is predicted by eqn (4.50) to be

$$(\mu_2^\gamma/\mu_0^\gamma)^{1/2} = (\mathcal{N}_b^\gamma/2)^{1/2}|h^\gamma| \tag{4.52}$$

which, therefore, takes the values $\sqrt{3/2}$, $\sqrt{2}$, $\sqrt{5/2}$, and $\sqrt{3}$ for $\gamma = l$, s, r, and t respectively (in units of $|h^\gamma|$). This checks with a direct summation over the square of the eigenvalues, $\varepsilon^\gamma$, in eqs (4.37)–(4.40). Note that for the particular case of $\alpha_h = \frac{1}{2}$ corresponding to $\lambda = 2$, all four eigenspectra have the same normalized second moment, $\mu_2^\gamma/\mu_0^\gamma$, since substituting eqn (4.42) into eqn (4.52) we have

$$\mu_2^\gamma/\mu_0^\gamma = h_0^2. \tag{4.53}$$

This is true for all the eigenspectra shown in Fig. 4.4 for the choice of $h^\gamma$ that is given by eqn (4.42) with $\lambda = 2$. It is a direct consequence of the structural energy difference theorem, since

$$\Delta U_{rep}^{\lambda=2} = 0 \Rightarrow \Delta(\mu_2/\mu_0) = 0 \tag{4.54}$$

as $\Phi_{rep} \alpha h^2 \alpha \mu_2$.

Thirdly, the skewness of the eigenspectrum is predicted by eqn (4.51) to be the sum over all three-membered rings within the molecule. Thus, the linear chain and the square have symmetric eigenspectra, since there are no three-membered rings and $\mu_3 = 0$. The rhombus, on the other hand, has

$$\mu_3^r = 2(4+2)(h^r)^3 = -12|h^r|^3, \tag{4.55}$$

whereas the tetrahedron has

$$\mu_3^t = 4(6)(h^t)^3 = -24|h^t|^3, \tag{4.56}$$

remembering that we can hop either clockwise or anticlockwise around a triangle. Hence, as is clear from Fig. 4.4, the eigenspectrum of the tetrahedron is more skewed than that of the rhombus. The eigenspectra are skewed downwards since $\mu_3 < 0$. Note that the eigenspectra in Fig. 4.4 have first been ordered from left to right according to whether the molecules are three-dimensional, two-dimensional, or one-dimensional respectively. For a given dimension they are then ordered from left to right according to their degree of skewness, $\hat{\mu}_3/\hat{\mu}_2^{3/2}$, where $\hat{\mu}_p = \mu_p/\mu_0$. For three-atom molecules, the triangle and linear chain have a degree of skewness of $-0.71$ and $0$ respectively. For four-atom molecules, the tetrahedron, rhombus, square, and linear chain have a degree of skewness of $-1.16$, $-0.76$, $0$, and $0$ respectively (which follows directly from eqs (4.52), (4.55), and (4.56)). For five-atom molecules, the trigonal bipyramid, monofinned tetrahedron, square pyramid, close-packed layer, pentagon, and linear chain have a degree of skewness of $-1.23$, $-1.05$, $-0.85$, $-0.76$, $0$, and $0$ respectively. For six-atom molecules the trigonal tripyramid, octahedron, pentagonal pyramid, trigonal prism, close-packed layer, hexagon, and linear chain have a degree of skewness of $-1.25$, $-1.00$, $-0.83$, $-0.39$, $-0.76$, $0$, and $0$ respectively.

Fourthly, counting all paths of length four within the linear chain, square, rhombus, and tetrahedron, we find

$$\mu_4^\gamma = \left.\begin{array}{c} 14 \\ 32 \\ 50 \\ 84 \end{array}\right\} |h^\gamma|^4 \quad \text{for } \gamma = \left.\begin{array}{c} 1 \\ s \\ r \\ t \end{array}\right\}. \tag{4.57}$$

The different types of paths are shown in Fig. 4.7 for the tetrahedron. We see that per starting atomic site there are 3 two-atom paths, 12 three-atom paths, and 6 four-membered ring paths, leading to a total of $4(3 + 12 + 6) = 84$ paths per tetrahedral molecule as in eqn (4.57). The fourth moment measures the unimodal versus bimodal behaviour of the

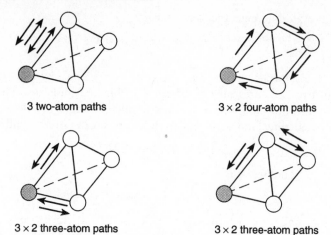

3 two-atom paths        $3 \times 2$ four-atom paths

$3 \times 2$ three-atom paths        $3 \times 2$ three-atom paths

**Fig. 4.7** Examples of nearest-neighbour paths of length four that contribute to the fourth moment of the tetrahedron's eigenspectrum. The solid atom indicates the site from which the path starts and to which it eventually returns. The number under each tetrahedron gives the total number of such paths starting and ending on the solid atom.

spectrum through a dimensionless shape parameter, $s$, namely

$$s = (\hat{\mu}_4/\hat{\mu}_2^2) - (\hat{\mu}_3^2/\hat{\mu}_2^3) - 1 \tag{4.58}$$

where $\hat{\mu}_p = \mu_p/\mu_0$. If $s < 1$, the spectrum is said to show bimodal behaviour, whereas if $s > 1$, it shows unimodal behaviour. The values of the normalized moments $\hat{\mu}_2$, $\hat{\mu}_3$, $\hat{\mu}_4$, and the shape parameters, $s$ are given in Table 4.3 for the linear chain, square, rhombus, and tetrahedron. We see that the tetrahedron shows perfect bimodal behaviour with $s = 0$, corresponding to the eigenspectrum in Fig. 4.4, splitting into a deep singly degenerate bonding level and a triply degenerate antibonding level. The square, on the other hand, lies exactly on the unimodal–bimodal borderline with $s = 1$, having

**Table 4.3** The normalized moments, $\hat{\mu}_2$, $\hat{\mu}_3$, and $\hat{\mu}_4$, and the shape parameter, $s$, for the linear chain, square, rhombus, and tetrahedron (cf eqn (4.58)).

| $\gamma$ | $\hat{\mu}_2$ | $\hat{\mu}_3$ | $\hat{\mu}_4$ | $s$ |
|---|---|---|---|---|
| t | 3 | −6 | 21 | 0 |
| r | 5/2 | −3 | 25/2 | 53/125 |
| s | 2 | 0 | 8 | 1 |
| l | 3/2 | 0 | 7/2 | 5/9 |

its sizeable bonding-antibonding level splitting neutralized by two non-bonding levels. The rhombus and linear chain take the values $s = 0.424$ and 0.555 respectively.

We are now in a position to understand why the bond energy curves in Figs 4.5 and 4.6 cross each other. Let us first consider the relative stability of the two three-atom molecules, the linear chain and the triangle. Their eigenspectra have identical first three moments $\mu_0$, $\mu_1$, and $\mu_2$ for the case of $\alpha_h = \frac{1}{2}$ corresponding to $\lambda = 2$ (cf eqn (4.54)), so that $\Delta\mu_p = 0$ for $p \leq 2$. However, the eigenspectrum of the triangle is skewed downwards due to a negative third moment resulting from the three-membered ring contributions, whereas the linear chain's eigenspectrum is symmetric. Thus $\Delta\mu_3 \neq 0$. This results in the two-bond energy curves crossing *once* as a function of electron count as is seen in the upper left-hand panel of Fig. 4.5.

Let us now consider the relative stability of the two four-atom clusters with symmetric eigenspectra, namely the linear chain and the square. Since $\mu_3 = 0$, they have identical first four moments $\mu_0$, $\mu_1$, $\mu_2$, and $\mu_3$ for $\alpha_h = \frac{1}{2}$, so that $\Delta\mu_p = 0$ for $p \leq 3$. However, we see from Table 4.3 that $\Delta\mu_4 \neq 0$; the linear chain's eigenspectrum shows marked bimodal behaviour compared to that of the square. This is reflected in the eigenspectra in Fig. 4.4: the linear chain has a distinct bonding-antibonding gap at the centre of the spectrum, whereas the square has a two-fold degenerate nonbonding level exactly at the centre. Moreover, this increased weight at the centre of the square's eigenspectrum is compensated for by a wider overall spectrum than that of the linear chain, in order that their second moments remain identical. Thus, as the electron occupancy increases, we expect to find the structural trend from square → linear chain → square; that is, the two bond energy curves for the square and the linear chain will cross *twice* as a function of electron count, as is indeed observed in the middle left-hand panel of Fig. 4.5.

In 1971 Ducastelle and Cyrot-Lackmann proved a very important moments theorem that captures this crossing behaviour of the bond energy curves. Their moments theorem states that if two eigenspectra have moments that are identical up to some level $p_0$, that is $\Delta\mu_p = 0$ for $p \leq p_0$, then the two bond energy curves must cross at least $(p_0 - 1)$ times as a function of electron count. Thus, we have seen in Fig. 4.5 that the triangle and the linear chain curves cross once, since $p_0 = 2$, whereas the square and the linear chain curves cross twice, since $p_0 = 3$.

We can, therefore, understand the origin of the four-atom structural trend from tetrahedron → rhombus → linear chain → square as a function of the electron count for the case of $\alpha_h = \frac{1}{2}$. The presence of three-membered rings in a given geometry skews the eigenspectrum asymmetrically downwards. Hence, the *close-packed* structures, the tetrahedron and the rhombus, are stabilized for fractional electron occupancies less than one-half but destabilized for fractional electron counts more than one half. On the other hand, the *open* structures, the square and the linear chain, have symmetric eigenspectra. In addition, the linear chain has fewer paths of length four than the square, so that its eigenspectrum is more bimodal. Hence, the linear chain is stabilized for a fractional electron occupancy around one-half but destabilized with respect to the square for fractional electron counts around three-quarters or greater. We see from the middle panel of Fig. 4.5 that as the degree of normalized hardness increases from $\alpha_h = \frac{1}{2}$, the open structures become destabilized with respect to the close-packed structures until for $\alpha_h = 1$, corresponding to a hard-core potential, the linear chain is no longer found anywhere as a ground-state structure. This general feature of moving from close-packed to open structures as the electron count increases is displayed by the $\mathcal{N} = 3$, 5, and 6 bond energy curves in Figs 4.5 and 4.6. We will see in Chapter 8 that the third moment skewing is also responsible for the observed trend from close-packed to open ground-state structure types across the sp-valent elements within the periodic table. Further examples of the application of moments to understanding structural trends within molecules can be found in Burdett (1985) and Lee (1991).

## 4.6. The bond order

The total bond energy of the molecule has been written as a sum over the occupied eigenvalues $\varepsilon_n$. The eigenvalues (and eigenfunctions) are a *global* property of the molecule that results from diagonalizing the LCAO or TB secular equation. In practice, when comparing the bonding in molecules and solids, we would like to define a *local* bond energy between individual pairs of atoms. This can be achieved by decomposing the energy as follows:

$$\mathcal{N} U_{\text{bond}} = \sum_n f_n \varepsilon_n = \sum_n f_n \int \psi^{(n)*} \hat{H} \psi^{(n)} \, d\mathbf{r} \qquad (4.59)$$

$$= \sum_{i,j} \left( \sum_n f_n c_i^{(n)*} c_j^{(n)} \right) \int \psi_i^* \hat{H} \psi_j \, d\mathbf{r} \qquad (4.60)$$

where the $c_i^{(n)}$ ($i = 1, 2, \ldots, \mathcal{N}$) are the components of the $n$th eigenfunction $\psi^{(n)}$ (cf eqn (4.31)). Hence, the total bond energy can be written as a sum over the individual bond energies as

$$\mathcal{N} U_{\text{bond}} = \tfrac{1}{2} \sum_{i,j}' U_{\text{bond}}^{ij}, \qquad (4.61)$$

where

$$U_{\text{bond}}^{ij} = -2|h(R_{ij})| \Theta_{ij} \qquad (4.62)$$

with

$$\Theta_{ij} = \sum_n f_n (c_i^{(n)*} c_j^{(n)} + c_j^{(n)*} c_i^{(n)})/2. \qquad (4.63)$$

The term $\Theta_{ij}$ is called the *bond order* between atoms $i$ and $j$.

The bond order has a simple physical interpretation. The eigenfunction,

$\psi^{(n)}$, can be written from eqn (4.31) as

$$\psi^{(n)} = \frac{1}{2}\left(\sum_i c_i^{(n)}\psi_i + \sum_j c_j^{(n)}\psi_j\right),$$

(4.64)

as $i$ and $j$ are dummy variables. Hence,

$$\psi^{(n)} = \frac{1}{2}\sum_{i,j}\left\{\frac{1}{\sqrt{2}}(c_i^{(n)} + c_j^{(n)})\left[\frac{1}{\sqrt{2}}(\psi_i + \psi_j)\right]\right.$$
$$\left. + \frac{1}{\sqrt{2}}(c_i^{(n)} - c_j^{(n)})\left[\frac{1}{\sqrt{2}}(\psi_i - \psi_j)\right]\right\}$$

(4.65)

Therefore, the total probability of locating an electron in the bonding state $(1/\sqrt{2})(\psi_i + \psi_j)$ compared to the antibonding state $(1/\sqrt{2})(\psi_i - \psi_j)$ will be given by

$$N_{ij}^+ - N_{ij}^- = \sum_n f_n\left\{\left|\frac{1}{\sqrt{2}}(c_i^{(n)} + c_j^{(n)})\right|^2 - \left|\frac{1}{\sqrt{2}}(c_i^{(n)} - c_j^{(n)})\right|^2\right\}$$

(4.66)

$$= 2\Theta_{ij}.$$

(4.67)

Thus, the bond order is one-half the difference between the number of electrons in the bonding state compared to the antibonding state. It takes its maximum value of unity when the bonding state is totally occupied with two electrons of opposite spin and the antibonding state has no electrons as, for example, in the hydrogen dimer.

Usually, however, the covalent bond is not *saturated* with a bond order of unity but is *unsaturated* with a bond order less than unity. This is illustrated by Fig. 4.8, which shows the bond orders for the linear chain, square, rhombus, and tetrahedron as a function of the electron count, $N$. The curves were evaluated from eqn (4.63) using the eigenfunctions of the TB secular equation that correspond to the eigenvalues eqs (4.37)–(4.40). The eigenfunctions for the square and the rhombus are given explicitly in Fig. 4.9. When the lowest bonding energy level of the square is occupied with two electrons of opposite spin, we see that the bond order between any pair of neighbouring atoms is given by $2(\frac{1}{2})(\frac{1}{2}) = 0.5$ as plotted in the upper right-hand panel of Fig. 4.8. Adding a further four electrons into the doubly degenerate nonbonding state contributes nothing to the bond order, since one or other of the neighbouring components of the eigenfunction vanishes. Finally, occupying the antibonding state reduces the bond order by $2(\frac{1}{2})(-\frac{1}{2}) = -0.5$, so that the bond order returns to zero at $N = 8$ when all the states are occupied.

The eigenspectrum of the rhombus is related to that of the square as shown in Fig. 4.9. The creation of three-membered ring contributions leads to the skewing of the eigenspectrum and the lifting of the degeneracy of the

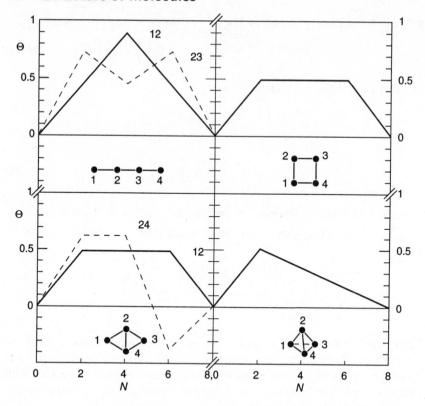

**Fig. 4.8** The bond order for the linear chain, square, rhombus, and tetrahedron as a function of the electron count, $N$.

nonbonding orbital. As can be seen from the nature of the two nonbonding eigenfunctions of the square, the degeneracy is lifted by one state moving away upwards due to the creation of an antibond across the rhombus, the other nonbonding state remaining unperturbed since it has no weight on the atoms terminating the shorter diagonal of the rhombus. The bond order of the outer bonds 12, 23, 34, and 41 is very similar to that of the square, its maximum value being slightly reduced from 0.5 to $2(0.557)(0.435) = 0.485$. The bond order of the inner bond, however, is dramatically different, dropping from its maximum value of $2(0.557)(0.557) = 0.620$ for $N = 2$ and 4 to a negative value of $-0.380$ for $N = 6$. This is due to occupation of a fully saturated antibonding state with a contribution to the bond order of $2[(1/\sqrt{2})(-1/\sqrt{2})] = -1$. From eqn (4.62) this corresponds to a repulsive rather than an attractive bond energy. Thus, whereas the sum of the global eigenvalues $\varepsilon_n$ is always negative for $0 < N < 8$, the local bond energies can be positive. Hence, a four-atom rhombic molecule with $N = 6$ would have

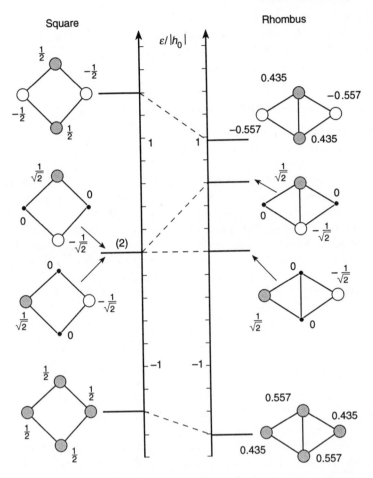

**Fig. 4.9** The eigenvalues and eigenfunctions of the square and rhombus. The two eigenspectra have identical first three moments $\mu_0$, $\mu_1$, and $\mu_2$.

its inner diagonal bond experiencing strong compressive forces from the outer bonds of the rhombus.

Figure 4.8 shows that the individual bonds in four-atom s-valent *clusters* are unsaturated, taking bond orders that are much less than unity. The two end bonds of the linear chain provide the only exception. This reduction in bond order compared to the isolated dimer is, of course, due to a given spherically symmetric s orbital forming bonds with all its neighbours. However, we will see later in Chapter 7 that the angular character of hybrid orbitals allows some sp-valent *solids* to exhibit saturated bond behaviour. Thus, the concept of the bond order is important because it not only quantifies the degree of saturated versus unsaturated character of the bond, but it also links the bonding of small molecules to that of extended solids. This latter theme has been explored recently in books by Hoffmann (1988), Sutton (1993), and Burdett (1995).

## 4.7. Linear versus bent triatomic molecules

The fact that $H_2O$ is a bent molecule but $CO_2$ is linear is directly related to the angular character of the valence orbitals which we have neglected so far in this chapter. We will first consider the simplest class of triatomic molecules, namely $AH_2$. It is known, for example, that $BeH_2$ is linear but that $BH_2$ like $OH_2$ is bent. Figure 4.10 shows the geometry of the molecule with bond angle $2\beta$ and our choice of coordinate axes with $O_x$ along the 2-fold rotational axis of the molecule and $O_z$ normal to the molecular plane. The hydrogen atoms are located at sites B and C respectively whereas the A atom is located at the origin, O. Since the molecular potential, $V_{ABC}$, is symmetric with respect to reflection in the $xz$ plane, the allowed molecular orbitals, $\psi$, must be either symmetric or antisymmetric across this plane. We, therefore, look for symmetric solutions

$$\psi^{(s)} = c_s^{(s)} \frac{1}{\sqrt{2}} (\psi_{B_s} + \psi_{C_s}) + c_p^{(s)}\psi_{A_x} \tag{4.68}$$

and asymmetric solutions

$$\psi^{(a)} = c_s^{(a)} \frac{1}{\sqrt{2}} (\psi_{B_s} - \psi_{C_s}) + c_p^{(a)}\psi_{A_y}. \tag{4.69}$$

We have assumed that the valence s state on the A atom lies sufficiently far below the valence p state that it may be treated as a nonbonding level (cf Fig. 2.16). The bonding will be taken to be between the $p_x$ and $p_y$ valence

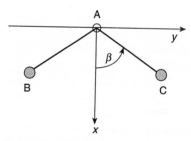

**Fig. 4.10** The geometry of the molecule $AH_2$ with bond angle $2\beta$. The two hydrogen atoms are located at sites B and C respectively, whereas atom A is located at the origin. The x-axis is along the 2-fold rotational axis of the molecule whereas the z-axis is normal to the molecular plane.

orbitals on the atom A and the 1s orbitals on the two hydrogen atoms. The $p_z$ orbital on atom A has not been included amongst the symmetric solutions in eqn (4.68) since it remains nonbonding due to the zero overlap between itself and the hydrogenic s orbitals.

The orthogonal TB secular equation corresponding to the symmetric solutions, therefore, takes the form

$$\begin{pmatrix} E_H - E & H_{sp} \\ H_{ps} & E_{A_p} - E \end{pmatrix} \begin{pmatrix} c_s^{(s)} \\ c_p^{(s)} \end{pmatrix} = 0, \tag{4.70}$$

where $E_H$ and $E_{A_p}$ are effective s and p on-site energy levels for hydrogen and atom A respectively. The Hamiltonian matrix element $H_{sp}$ is given by

$$H_{sp} = \frac{1}{\sqrt{2}} \int (\psi_{B_s} + \psi_{C_s}) \hat{H} \psi_{A_x} \, d\mathbf{r}. \tag{4.71}$$

But choosing new axes $x'$, $y'$, $z'$, so that $Ox'$ lies along AC, we have from Fig. 4.11 where $\theta = \beta$, that

$$\int \psi_{C_s} \hat{H} \psi_{A_x} \, d\mathbf{r} = \int \psi_{C_s} \hat{H} [\cos \beta \psi_{A_{x'}} - \sin \beta \psi_{A_{y'}}] \, d\mathbf{r} = -sp\sigma \cos \beta, \tag{4.72}$$

since $\int \psi_{C_s} \hat{H} \psi_{A_{y'}} \, d\mathbf{r} = 0$ by symmetry. Moreover, because $xz$ is a mirror plane, replacing $\psi_{C_s}$ by $\psi_{B_s}$ in eqn (4.72) leaves the integral unchanged. Hence, from eqs (4.71) and (4.72),

$$H_{sp} = -\sqrt{2} sp\sigma \cos \beta. \tag{4.73}$$

The $2 \times 2$ TB secular equation may now be solved to yield the eigenvalues

$$E^{(s)} = \bar{E} \pm \tfrac{1}{2}[(\Delta E)^2 + 8sp\sigma^2 \cos^2 \beta]^{1/2} \tag{4.74}$$

where $\bar{E} = \tfrac{1}{2}(E_H + E_{A_p})$ and $\Delta E = E_H - E_{A_p}$. Similarly, the eigenvalues for the asymmetric solutions are given by

$$E^{(a)} = \bar{E} \pm \tfrac{1}{2}[(\Delta E)^2 + 8sp\sigma^2 \sin^2 \beta]^{1/2} \tag{4.75}$$

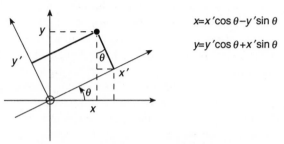

$x = x' \cos \theta - y' \sin \theta$

$y = y' \cos \theta + x' \sin \theta$

**Fig. 4.11** The transformation of $(x, y)$ into $(x', y')$ by a rotation of the axes through $\theta$.

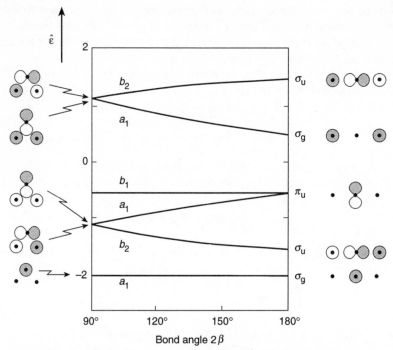

**Fig. 4.12** The normalized eigenvalues, $\hat{\varepsilon}$, of the triatomic molecule $AH_2$ as a function of the bond angle, $2\beta$, for the particular choice of normalized atomic energy level mismatch, $\delta = 1$. The symmetries of the eigenfunctions are also shown for the bent and linear configurations.

The energy may be measured with respect to the average energy level $\bar{E}$ and normalized by the bond integral $sp\sigma$, so that

$$\hat{\varepsilon}^{(s)} = (E^{(s)} - \bar{E})/sp\sigma = \pm\tfrac{1}{2}(\delta^2 + 8\cos^2\beta)^{1/2} \qquad (4.76)$$

and

$$\hat{\varepsilon}^{(a)} = (E^{(a)} - \bar{E})/sp\sigma = \pm\tfrac{1}{2}(\delta^2 + 8\sin^2\beta)^{1/2} \qquad (4.77)$$

where $\delta = \Delta E/sp\sigma$. These normalized eigenvalues are plotted in Fig. 4.12 for the particular choice of normalized atomic energy-level mismatch $\delta = 1$. Also drawn are the nonbonding $\psi_{A_s}$ and $\psi_{A_z}$ levels assuming $(E_{A_s} - \bar{E})/sp\sigma = -2$ for the former. For the linear molecule with a bond angle of $180°$, the bonding and antibonding states are labelled, $\sigma_u$, since their eigenfunctions are odd with respect to inversion about the centre of the molecule. The two nonbonding s states are labelled, $\sigma_g$ (corresponding to the low-lying s levels on atom A with energy $\hat{\varepsilon}_{A_s} = -2$ and the even combination of the two hydrogenic 1s orbitals with energy $\hat{\varepsilon}_H = \delta/2$). The nonbonding p states are labelled $\pi_u$ (corresponding to the $p_x$ and $p_z$ orbitals with their odd inversion symmetry with energy $\hat{\varepsilon}_{A_p} = -\delta/2$).

For the bent molecules the states are labelled $a_1$, $b_1$ or $b_2$ according to their symmetry. The state, $a_1$, is unchanged by a $180°$ rotation about the 2-fold rotational axis along $Ox$ or by reflection in either of the two mirror planes $xz$ or $xy$ respectively. The state, $b_1$, is unchanged by reflection in $xz$ but is changed by a $180°$ rotation about $Ox$ or reflection in $xy$. The state, $b_2$, is unchanged by reflection in $xy$ and is changed by a $180°$ rotation about $Ox$ or reflection in $xz$. Thus, the most bonding and antibonding levels in Fig. 4.12 are denoted by the symmetry label, $b_2$, whereas the least bonding and antibonding levels are denoted by the symmetry label, $a_1$. The non-bonding $p_z$ orbital has $b_1$ symmetry.

Figure 4.12 correlates the energy levels in the bent molecule with those in the linear molecule. This correlation diagram or Walsh diagram allows us to understand the origin of the change from linear to bent behaviour that is observed amongst the $AH_2$ trimers as the valence electron count increases. $BeH_2$ has four valence electrons, so that the two lowest-lying molecular states $a_1$ and $b_2$ are both doubly occupied with a pair of opposite spin electrons. We see that, whereas the energy of the lowest lying state $a_1$ is essentially independent of bond angle, the energy of state $b_2$ becomes more bonding as the bond angle increases due to the increased overlap between the central $p_y$ orbital and the outer hydrogenic s orbitals. Thus, the ground state of $BeH_2$ takes the linear configuration corresponding to maximum overlap for the $b_2/\sigma_u$ state.

$CH_2$, on the other hand, has six valence electrons, so that it takes the ground-state singlet configuration $(a_1)^2(b_2)^2(a_1)^2$. It follows from eqs (4.76) and (4.77) that the normalized bond angle-dependent contribution to the energy can be written

$$2(\hat{\varepsilon}^{(a)} + \hat{\varepsilon}^{(s)}) = -[(\delta^2 + 8\cos^2\beta)^{1/2} + (\delta^2 + 8\sin^2\beta)^{1/2}], \qquad (4.78)$$

which has its minimum value for the bond angle $2\beta = 90°$. This is consistent with Fig. 4.12 where we see that the average value of the $\sigma_u$ and $\pi_u$ levels of the linear molecule is $-1.000$, whereas the degenerate $a_1/b_2$ level for $2\beta = 90°$ is $-1.118$. Thus, $CH_2$ will be a bent molecule. The structure of $BH_2$ with one less valence electron than that of $CH_2$ is not so clear cut as the simple TB model will predict a bent molecule for $|\delta| < \sqrt{8/3}$ but a linear molecule for $|\delta| > \sqrt{8/3}$. In practice, the ground state of $BH_2$ is found to be bent but the excited state $(a_1)^2(b_2)^2(b_1)^1$ is linear or nearly linear. As is clear from Fig. 4.12, $NH_2$ and $OH_2$, with seven and eight valence electrons respectively, will be bent. The fact that $H_2O$ has a bond angle of $104°$ and not the predicted $90°$ is due partly to our neglect of any sp hybridization or mixing on the oxygen site (cf. question 4.4 in the Problems section at the end of the book). Molecules $H_2S$, $H_2Se$, and $H_2Te$ do, however, take the bond angles of $92°$, $91°$, and $90°$ respectively.

The structural trend from linear to bent (to linear) as the energy levels are filled progressively with electrons reflects the behaviour of the fourth

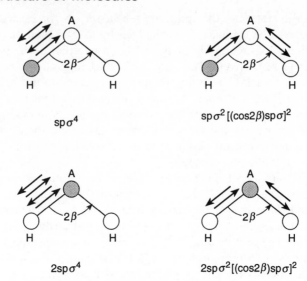

**Fig. 4.13** Examples of nearest-neighbour paths of length four that contribute to the fourth moment of the AH$_2$ eigenspectrum for the case of vanishing atomic energy-level mismatch. The solid atom indicates the site from which the path starts and to which it eventually returns. The number prefactor under each trimer gives the total number of such paths starting and ending on the solid atom. The three-atom paths involve the square of cos $2\beta$ due to a reduction in magnitude of the p$_\sigma$ orbital on rotation through $2\beta$ as in Fig. 4.11.

moment. We see from Fig. 4.13 that the fourth moment comprises both two-atom and three-atom contributions. The two-atom contributions are of the type sp$\sigma^4$, whereas the three-atom contributions vary with the bond angle as sp$\sigma^4 \cos^2 2\beta$. The angular factor in the latter reflects the cos $2\beta$ reduction in strength of the p$_\sigma$ orbital on rotating it through $2\beta$ as shown in Fig. 4.11. The total fourth moment (for the case of $\delta = 0$, for simplicity) is, therefore, given by

$$\mu_4 = 4(1 + \cos^2 2\beta)\text{sp}\sigma^4, \tag{4.79}$$

which can be checked by evaluating $2[(\hat{\varepsilon}^{(s)})^4 + (\hat{\varepsilon}^{(a)})^4]$ from eqs (4.76) and (4.77). The second moment $\mu_2$, on the other hand, involves only self-returning paths between nearest neighbours and is bond angle independent, namely $\mu_2 = 4\text{sp}\sigma^2$.

Thus, the dimensionless shape parameter in eqn (4.58) takes the value $s = \cos^2 2\beta$. Therefore, the bent molecule with a bond angle of 90° has a perfectly bimodal eigenspectrum corresponding to $s = 0$, as is reflected by its two degenerate levels in Fig. 4.12. The linear molecule with a bond angle of 180° takes a value of $s = 1$ and lies on the bimodal–unimodal borderline as is reflected by its four evenly spaced levels in Fig. 4.12. Hence, the bent

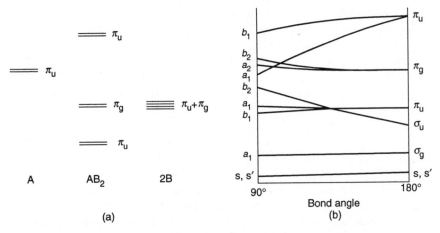

**Fig. 4.14** (a) The energy levels of the $\pi$ states for a single free atom A, two non-interacting atoms B, and the triatomic molecule $AB_2$. (b) The Walsh diagram for the $AB_2$ molecule. (After Williams (1979).)

molecule will be stabilized for fractional occupancies around one-half, the linear molecule for fractional occupancies away from one-half. This demonstrates the importance of the angular dependence of the bond integrals in structural determination. Spherically symmetric s orbitals would have bond angle independent moments, thereby providing no differentiation (within a first nearest-neighbour model) between the bent and linear geometries considered above (for $2\beta > 60°$).

The Walsh diagram for $AB_2$ molecules such as $CO_2$ or $NO_2$ is shown in Fig. 4.14(b). It is similar to Fig. 4.12 for the $AH_2$ trimers except for the presence of extra $\pi$-related levels that result from the valence p orbitals on the two B sites that were, of course, absent from the two hydrogen sites. In addition, the very deep valence s state on the two B sites is also sketched and labelled s, s'. The p states are drawn in Fig. 4.14(a) to be at a lower level on the B site than on the central A site, since this is expected from the ordering of the atomic energy levels in C and O, for example (cf Fig. 2.16). We have already seen in Fig. 3.13 that the $p_x$ and $p_z$ orbitals on the two B sites can be combined in either $\pi_u$ or $\pi_g$ symmetry that is consistent with a linear geometry. These p levels then split in the $AB_2$ linear molecule as indicated in the middle panel of Fig. 4.14(a) with the bonding $\pi_u$ wave function being weighted mainly on site B, the antibonding $\pi_u$ wave function mainly on site A.

We can now understand why $CO_2$ is observed to be linear but $NO_2$ to be bent. From Fig. 4.14(b) the sixteen valence electrons in $CO_2$ will take the electronic configuration $(s)^2(s')^2(a_1)^2(b_1)^2(a_1)^2(b_2)^2(a_2)^2(b_2)^2$. Due to the marked increase in bonding of the $b_2/\sigma_u$ state with increasing bond angle, the ground state of $CO_2$ will be linear. On the other hand, the additional valence electron in $NO_2$ occupies the $a_1/\pi_u$ state whose binding energy decreases strongly with increasing bond angle, so that $NO_2$ will be bent. $CO_2$ and $NO_2^+$ are both observed to take a bond angle of $180°$, whereas $CO_2^-$ and $NO_2$ are found with bond angles of $134°$ and $127°$ respectively. As expected from the Walsh diagram, increasing the number of valence electrons further reduces the bond angle. Thus, $NO_2^-$, $O_3$, $SO_2$, and $CF_2$ with eighteen valence electrons have bond angles of $115°$, $117°$, $120°$, and $105°$ respectively.

We should note, however, that the direct correlation between the electron count N and molecular structure is not always found amongst the sp-valent trimers. For example, although $BeF_2$, $MgF_2$, $ZnF_2$, $CdF_2$, and $HgF_2$ are linear as expected for $N = 16$, $CaF_2$, $SrF_2$, and $BaF_2$ are bent. The latter are, therefore, located in the bent domain at the upper left-hand corner of the $AB_2$ structure map in Fig. 1.14. As we have already seen in our comparison of the $Si_2$ and $C_2$ energy levels in Fig. 3.13, the exact ordering of the states can be sensitive to the particular elements under consideration. The alkaline earths Ca, Sr, and Ba fall at the beginning of the transition metal series. They, therefore, contain unoccupied d states that influence the relative hierarchy of levels within the conventional Walsh diagram for the sp-valent triatomic molecules. We will see in the next chapter that the presence of these d states dramatically changes the shape of the electronic density of states of the alkaline earth metals Ca, Sr, and Ba compared to the isovalent metals Mg, Zn, and Cd.

# References

Burdett, J. K. (1988). *Accounts of Chemical Research* **21**, 189.

Burdett, J. K. (1995). *Chemical bonding in solids*, Oxford University Press, New York.

Cyrot-Lackmann, F. (1967). *Advances in Physics* **16**, 393.

Ducastelle, F. and Cyrot-Lackmann, F. (1971). *Journal of Physics and Chemistry of Solids* **32**, 285.

Hoffmann, R. (1988). *Solids and surfaces: a chemist's view of bonding in extended structures*, VCH, New York.

Lee, S. (1991). *Accounts of Chemical Research* **24**, 249.

Pettifor, D. G. (1986). *Journal of Physics* **C19**, 285.

Shah, M. and Pettifor, D. G. (1993). *Journal of Alloys and Compounds* **197**, 145.

Sutton, A. P. (1993). *Electronic structure of materials*, Oxford University Press.

Wang, Y., George, T.F., Lindsay, D. M., and Beri, A. C. (1987). *Journal of Chemical Physics* **86**, 3493.

Williams, A. F. (1979). *A theoretical approach to inorganic chemistry*, Springer-Verlag, Berlin.

# 5
# Bonding of sp-valent metals

## 5.1. Introduction

A metal is often pictured as a gas of free electrons into which has been immersed a lattice of positive ions. The metallic bond is, therefore, thought of as having no directional character in contrast to saturated covalent bonds with their hybrid orbitals resistant to bond bending. The metal atoms behave like hard spheres that are held together by the all-pervasive electron glue, taking close-packed structures such as fcc, bcc, or hcp. Unlike valence compounds with their restrictive requirement of electron-pair bonding between neighbouring sites, metals can form alloys over a wide range of composition, atoms of one type replacing those of another with comparative ease within the electron gas.

We will see in this chapter that this conventional view of the metallic bond is indeed an excellent description of sp-valent metals such as sodium, magnesium, and aluminium. We will begin, therefore, by linking the world of small molecules and extended solids together by applying the jellium model to a study of cohesion in atomic clusters. We will find that it predicts special stability for alkali metal clusters containing *magic numbers* of atoms that correspond to electronic shell closings. However, as we have already seen in Chapter 2, jellium is only in equilibrium at one specific electron density. The underlying ionic lattice is required for differentiating between the elements and predicting the properties of the sp-valent metals. The influence of a periodic crystalline potential on the electronic structure is introduced through the one-dimensional Kronig–Penney model that illustrates all the essential features of band theory. The key concept of the pseudopotential is presented in order to account for the well-known but surprising fact that the free-electron gas is only very weakly perturbed by the ionic lattice. This allows us to develop a quantitative nearly free electron (NFE) model of the metallic bond that is entirely consistent with the usual picture of a metal as a gas of free electrons into which has been immersed a lattice of positive ions.

## 5.2. Jellium: from small molecules to the bulk

The jellium model of the free-electron gas can account for the increased abundance of alkali metal clusters of a certain size which are observed in mass spectroscopy experiments. This occurrence of so-called magic numbers is related directly to the electronic shell structure of the atomic clusters. Rather than solving the Schrödinger equation self-consistently for jellium clusters, we first consider the two simpler problems of a free-electron gas that is confined either within a sphere of radius, $R$, or within a cubic box of edge length, $L$ (cf. problem 28 of Sutton (1993)). This corresponds to imposing hard-wall boundary conditions on the electrons, namely

$$\psi(\mathbf{r}) = 0 \quad \text{for} \quad \begin{cases} r = R & \text{(sphere)} \\ x = \pm L/2, \; y = \pm L/2, \; z = \pm L/2 & \text{(cube)}. \end{cases} \quad (5.1)$$

The sphere radius, $R$, and edge-length, $L$, may be written in terms of the radius of the sphere containing one electron, $r_s$, and the number of monovalent atoms in the cluster $\mathcal{N}$, as

$$\tfrac{4}{3}\pi R^3 = L^3 = \tfrac{4}{3}\pi r_s^3 \mathcal{N}. \quad (5.2)$$

The eigenfunctions of the free-electron Schrödinger equation with spherical boundary conditions can be written in separable form like that for the hydrogen atom, eqn (2.48), namely

$$\psi(\mathbf{r}) = A j_l(\kappa r) Y_l^m(\theta, \phi), \quad (5.3)$$

where $A$ is a normalization constant, $\kappa$ is related to the eigenvalue $E$ through $E = (\hbar^2/2m)\kappa^2$, and $j_l(\kappa r)$ is a spherical Bessel function (see, for example, Gasiorowicz (1974)). This latter function is oscillatory, as is evident from the s- and p-related functions which are given by

$$j_0(\kappa r) = (\sin \kappa r)/\kappa r \quad (5.4)$$

and

$$j_1(\kappa r) = (\sin \kappa r)/\kappa^2 r^2 - (\cos \kappa r)/\kappa r. \quad (5.5)$$

The boundary condition eqn (5.1) then determines the eigenvalues $E_{nl}$ since, for example, for $l = 0$ we must have $\sin \kappa R = 0$, so that $\kappa_{ns} = n\pi/R$, where $n$ is a positive integer. The eleven lowest roots of $j_l(\kappa R) = 0$ are given in Table 5.1. We see that the three roots corresponding to $l = 0$ are $\kappa R = \pi$, $2\pi$, and $3\pi$ as expected. Note that $l = 4$ and $l = 5$ states are referred to as g and h respectively following as they do after the f states corresponding to $l = 3$.

The eigenspectrum of the free-electron gas confined within a sphere of

**Table 5.1** The eleven lowest roots $\kappa_{nl}R$ of $j_l(\kappa R) = 0$

|         | $l = 0$ | $l = 1$ | $l = 2$ | $l = 3$ | $l = 4$ | $l = 5$ |
|---------|---------|---------|---------|---------|---------|---------|
| $n = 1$ | 3.14    | 4.49    | 5.76    | 6.99    | 8.18    | 9.36    |
| $n = 2$ | 6.28    | 7.73    | 9.10    | 10.42   |         |         |
| $n = 3$ | 9.42    |         |         |         |         |         |

radius, $R$, is, therefore, given by

$$E_{nl} = \frac{\hbar^2}{2m} \kappa_{nl}^2 = \frac{\hbar^2}{2mR^2} (\kappa_{nl}R)^2. \tag{5.6}$$

This may be compared with that for a free-electron gas confined within a cube of side, $L$, namely

$$E_n = \frac{\hbar^2 \pi^2}{2mL^2} (n_x^2 + n_y^2 + n_z^2), \tag{5.7}$$

where $n = (n_x, n_y, n_z)$ and $n_x, n_y, n_z \geq 1$. This result follows directly from imposing infinite-barrier boundary conditions on the cube rather than the periodic boundary conditions which we considered earlier in section 2.5. Since $R$ and $L$ are related through eqn (5.2), the eigenvalues of the cube and the sphere may be plotted on the same normalized energy scale $(2mL^2/\hbar^2\pi^2)E$, as displayed in Fig. 5.1. We see that the electronic structure of the spherical free-electron gas shows shell closing for the magic numbers 2, 8, 18, 20, 34, 40, 58, 68, 90, 92, 106, ..., whereas the cubic free-electron gas displays the many more magic numbers 2, 8, 14, 20, 22, 34, 40, 46, 52, 64, 70, 76, 88, 96, 108, ... respectively. The former ordering is the same as that predicted by the simplest shell model of the nucleus in which the protons and neutrons are confined by a spherical hard-wall barrier.

We expect the sphere to be more stable than the cube, since it has a 20% smaller surface area and hence less surface energy. This is indeed the case. The average kinetic energy of a sphere containing $\mathcal{N}$ monovalent atoms can be written from eqs (5.6) and (5.2) as

$$U_{ke}^{sphere}(\mathcal{N}) = \left\{ 0.452 \left[ \sum_{nl} (\kappa_{nl}R)^2 f_{nl} / \mathcal{N}^{5/3} \right] \right\} U_{ke}(\infty), \tag{5.8}$$

where $f_{nl}$ is the electronic occupancy of energy level $E_{nl}$. The term $U_{ke}(\infty)$ is the average kinetic energy per electron of an infinite free-electron gas, which from eqn (2.44) is given by

$$U_{ke}(\infty) = \frac{\hbar^2}{2m} \frac{2.210}{r_s^2}. \tag{5.9}$$

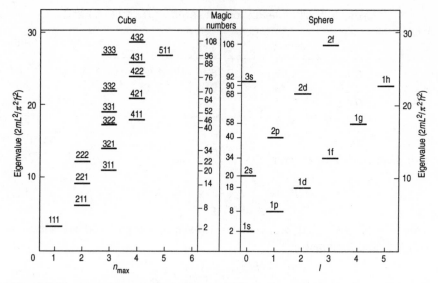

**Fig. 5.1** The eigenspectrum of a free-electron gas confined to a cube (left-hand panel) or a sphere (right-hand panel). The magic numbers corresponding to shell closings are given in the middle panel.

Similarly, the average kinetic energy of a cube containing $\mathcal{N}$ monovalent atoms can be written from eqs (5.7) and (5.2) as

$$U_{\mathrm{ke}}^{\mathrm{cube}}(\mathcal{N}) = \left\{ 1.719 \left[ \sum_{n} (n_x^2 + n_y^2 + n_z^2) f_n \right] \middle/ \mathcal{N}^{5/3} \right\} U_{\mathrm{ke}}(\infty), \quad (5.10)$$

where $f_n$ is the electronic occupancy of energy level, $n = (n_x, n_y, n_z)$. The infinite limit for the sphere and the cube are identical, since the ratio of surface to bulk atoms varies as $\mathcal{N}^{-1/3}$, which vanishes in the limit of infinite $\mathcal{N}$.

Figure 5.2 plots the average kinetic energy as a function of the number of monovalent atoms, $\mathcal{N}$, for the cube and sphere respectively. We see that the sphere is everywhere more stable than the cube as expected. Interestingly, however, even for the sizeable 100-atom cluster, the kinetic energy has still not yet fallen to the infinite free-electron gas value, being some 40% larger. For metallic sodium with an average kinetic energy of $3/5E_F = 1.9$ eV, this implies that the Na$_{100}$ atomic cluster would be about 0.8 eV per atom less stable than the bulk.

In practice, this large difference is reduced by about one-quarter because the boundary condition is not infinitely hard as implied by eqn (5.1) but much softer. Within the jellium model the positive ions are smeared out uniformly within a sphere of radius, $R$, so that there is indeed an abrupt discontinuity in background charge density at $r = R$. However, the free-

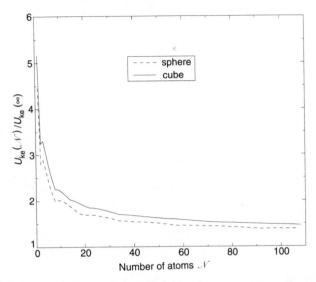

**Fig. 5.2** The ratio of the kinetic energy of an $\mathcal{N}$ monovalent atom cluster to that of the infinite bulk as a function of $\mathcal{N}$ for cubic (solid curve) and spherical (dashed curve) boundary conditions.

electron gas is free to spill out across this spherical boundary into the attractive region outside where the ionic potential falls off inversely with distance. Exactly how far the electrons 'spill out' and by how much their energy is lowered compared to the hard-boundary case requires the numerical self-consistent solution of the Schrödinger equation. Figure 5.3 shows the computed binding energy per atom for sodium clusters (assuming $r_s = 4.0$ au) as a function of the number of atoms $\mathcal{N}$. An infinite sample of jellium would have a binding energy per atom of 2.2 eV approximately (cf Fig. 2.11). Thus, we see that the binding energy of the $Na_{100}$ atomic cluster is within about 10% of that of bulk jellium.

The simple hard-wall boundary condition, eqn (5.1), does yield the correct ordering of the energy levels as shown in Fig. 5.1. The only exceptions are the 3s and 1h states, which have their sequence reversed compared to that of the self-consistent jellium predictions. Experimentally, the most frequently occurring sodium clusters are indeed $Na_8$ and $Na_{20}$, as expected from their special stability in Fig. 5.3.

## 5.3. General principles of band theory

The variation in equilibrium bulk properties between one sp-valent metal and the next cannot be understood within the jellium model, since it has obscured the chemical behaviour of the elements by smearing out the ion

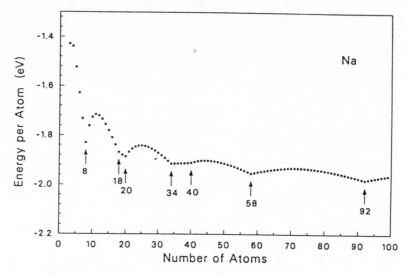

Fig. 5.3 The total energy per atom of sodium clusters versus the number of atoms in the cluster, evaluated within the self-consistent jellium model. (From Cohen (1987).)

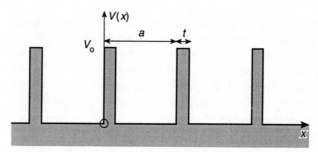

Fig. 5.4 The one-dimensional Kronig–Penney potential.

cores into a uniform positive background. In practice, the ionic lattice introduces an attractive periodic potential, which scatters the electrons as they travel through the crystal, thereby resulting in the band gaps that open up in the free-electron eigenspectrum. In this section we will illustrate the principles of band theory by considering the one-dimensional Kronig–Penney model. This model is sufficiently general that it spans all the way from the nearly free electron (NFE) regime (in which the band gaps are small and the electronic states are free-electron-like) to the tight binding (TB) regime (in which the band gaps are large and the electronic states are free-atom-like).

Figure 5.4 shows the one-dimensional potential $V(x)$ of the Kronig–Penney model, which comprises square wells that are separated by barriers of height,

$V_0$, and thickness t. It is periodic with repeat distance $a$ so that

$$V(x + na) = V(x) \qquad (5.11)$$

where $n$ is an integer. Because all the wells are equivalent, the probability of locating the electron in a given well must be the same for all wells, so that

$$|\psi(x + na)|^2 = |\psi(x)|^2. \qquad (5.12)$$

For $n = 1$ this implies that

$$\psi(x + a) = e^{ika}\psi(x), \qquad (5.13)$$

where $k$ is a number (in units of $1/a$) which specifies the phase factor $e^{ika}$ linking the wave functions in neighbouring wells. Repeating eqn (5.13) $n$ times gives

$$\psi_k(x + na) = e^{ikna}\psi_k(x) \qquad (5.14)$$

which is the usual statement of *Bloch's theorem* in one dimension. Thus, the translational symmetry of the lattice leads to the eigenfunctions being characterized by the Bloch vector $k$. It is only defined modulo $(2\pi/a)$ since $k + m(2\pi/a)$ results in the same phase factor in eqn (5.14) as $k$ alone. It is, therefore, customary to label the wave function $\psi_k$ by restricting $k$ to lie within the first *Brillouin zone* that is defined by

$$-\pi/a \le k \le +\pi/a. \qquad (5.15)$$

We note that in one dimension $na$ is a *direct* lattice vector, whereas $m(2\pi/a)$ is a *reciprocal* lattice vector. Their product is an integral multiple of $2\pi$.

Bloch's theorem enables us to solve the infinite one-dimensional problem directly by looking for the solution $\psi_k(x)$ in a given square well and then repeating it periodically through the lattice using eqn (5.14). Consider, therefore, the region of the first square well $0 \le x \le a$. Inside the well where the potential vanishes for $t < x < a$ the solution can be written as a linear combination of plane waves travelling to the right and to the left, namely

$$\psi_k(x) = A\, e^{iKx} + B\, e^{-iKx} \qquad (5.16)$$

where

$$K = (2mE/\hbar^2)^{1/2} \qquad (5.17)$$

with $E$ the eigenvalue corresponding to eigenfunction, $\psi_k$. We note that $K$ is a direct measure of the energy of the electron $E$, being directly proportional to its square root $E^{1/2}$. Under the potential barrier $V_0$ for $0 < x < t$ the solution can be written as a linear combination of exponential functions, namely

$$\psi_k(x) = C\, e^{\kappa x} + D\, e^{-\kappa x}, \qquad (5.18)$$

where

$$\kappa = [2m(V_0 - E)/\hbar^2]^{1/2}. \qquad (5.19)$$

The coefficients $A$, $B$, $C$, and $D$ are found in the usual manner by matching $\psi_k$ and its first derivative, $\psi'_k$ across the boundaries at $x = t$ and $x = a$ respectively. Matching at $x = t$ using eqs (5.16) and (5.18) we have

$$A\, e^{iKt} + B\, e^{-iKt} = C\, e^{\kappa t} + D\, e^{-\kappa t} \qquad (5.20)$$

and

$$iK(A\, e^{iKt} - B\, e^{-iKt}) = \kappa(C\, e^{\kappa t} - D\, e^{-\kappa t}). \qquad (5.21)$$

Matching at $x = a$ using Bloch's theorem to provide the solution for $a < x < a + t$ in terms of the solution eqn (5.18) for $0 < x < t$, we have

$$A\, e^{iKa} + B\, e^{-iKa} = e^{ika}(C + D) \qquad (5.22)$$

and

$$iK(A\, e^{iKa} - B\, e^{-iKa}) = \kappa\, e^{ika}(C - D). \qquad (5.23)$$

These four equations, eqs (5.20)–(5.23), have a solution only if the determinant of the coefficients of $A$, $B$, $C$, and $D$ vanishes. After non-trivial determinantal manipulation we find

$$[(\kappa^2 - K^2)/2\kappa K]\sinh \kappa t \sin K(a - t) + \cosh \kappa t \cos K(a - t) = \cos ka \qquad (5.24)$$

This equation may be simplified by considering the limit in which the barrier thickness becomes increasingly thin (i.e. $t \to 0$) but the barrier height becomes increasingly high (i.e. $V_0 \to \infty$) in such a way that the area under the barrier remains constant, that is

$$V_0 t = \text{constant} = (\hbar^2/ma)\mu. \qquad (5.25)$$

The parameter, $\mu$, measures the strength of the Kronig–Penney barrier between neighbouring square wells. In this limit $\kappa \sim (2mV_0/\hbar^2)^{1/2} \to \infty$, whereas $\kappa t \sim (2mV_0 t^2/\hbar^2)^{1/2} \to 0$. Thus, substituting into eqn (5.24) and using the small argument Maclaurin expansions, $\sinh \kappa t \approx \kappa t$ and $\cosh \kappa t \approx 1$, we have

$$\cos Ka + \mu \sin Ka/Ka = \cos ka, \qquad (5.26)$$

which links the energy, $E = (\hbar^2/2m)K^2$, to the Bloch vector, $k$.

Bloch-like solutions or travelling waves, therefore, only exist for those values of $K$, and hence energy $E$, for which the magnitude of the left-hand side of eqn (5.26) is less than or equal to unity, since $|\cos ka| \le 1$. Thus, as illustrated by Fig. 5.5(a), energy gaps open up in the spectrum as travelling solutions are only found between a and b, c and d, e and f, g and h, etc. The plot, Fig. 5.5(b), of the eigenvalues, $E(k)$, as a function of the Bloch vector, $k$, within the first Brillouin zone is called the *band structure*. It accounts naturally for the division of materials into metals (when the uppermost occupied band is partially filled) and semiconductors or insulators (when the uppermost occupied band is totally filled with a small or large gap to the next unoccupied band of states).

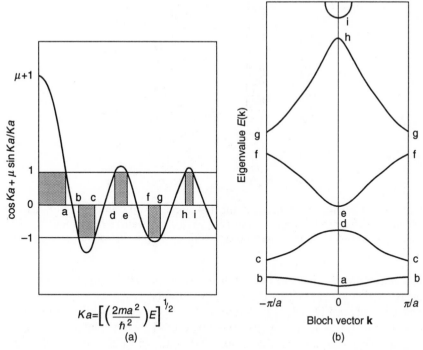

$$Ka=\left[\left(\frac{2ma^2}{\hbar^2}\right)E\right]^{1/2}$$

(a)

(b)

**Fig. 5.5** (a) Plot of the function $\cos Ka + \mu \sin Ka/Ka$ that appears on the left-hand side of eqn (5.26). Travelling wave or Bloch-like solutions are forbidden for those ranges of $Ka$ for which the magnitude of the function is greater than unity as shown by the shaded regions. (b) The resultant band structure $E$ versus $k$.

The Kronig–Penney band structure can range from free-electron-like behaviour to free-atom-like behaviour by changing the strength of the barrier, $\mu$. It follows from eqn (5.26) that as $\mu \to 0$, $\cos Ka \to \cos ka$, and hence

$$E = (\hbar^2/2m)K^2 \to (\hbar^2/2m)[k + n(2\pi/a)]^2, \tag{5.27}$$

which is, of course, the free-electron result (where the energy levels have been 'folded back' into the first Brillouin zone, $-\pi/a \le k \le \pi/a$). On the other hand as $\mu \to \infty$,

$$\sin Ka = Ka(\cos ka - \cos Ka)/\mu \to 0, \tag{5.28}$$

so that

$$E = (\hbar^2/2m)K^2 \to (\hbar^2/2m)(n\pi/a)^2, \tag{5.29}$$

which is, of course, the result for electrons confined within a one-dimensional hard-walled box (cf eqn (5.7)). In this latter case the band structure shows no dispersion, since $E(k) = $ constant, corresponding to a discrete

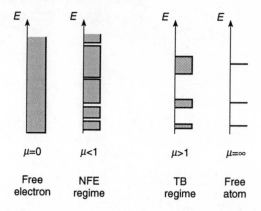

Fig. 5.6 Schematic illustration of the behaviour of the energy bands as the strength of the Kronig–Penney barrier $\mu$ varies between zero and infinity. The so-called free-atom limit in fact corresponds to a free or isolated square well.

eigenspectrum like that of a free atom. This change in behaviour of the energy bands, which takes place as the strength of the barrier $\mu$ varies is illustrated schematically in Fig. 5.6. This chapter and the next are concerned with those sp-valent metals that are well described by the NFE approximation, whereas the last two chapters of the book deal with sp-valent semiconductors, d-valent transition metals, and pd-valent intermetallics, which are well described by the TB approximation.

These fundamental ideas of band theory can be extended to three dimensions. In particular, Bloch's theorem takes the form

$$\psi_k(\mathbf{r} + \mathbf{R}) = e^{i\mathbf{k} \cdot \mathbf{R}} \psi_k(\mathbf{r}) \tag{5.30}$$

where $\mathbf{R}$ is any direct lattice vector which may be expressed in terms of the fundamental translation vectors $\mathbf{a}_1$, $\mathbf{a}_2$, and $\mathbf{a}_3$ as

$$\mathbf{R} = n_1\mathbf{a}_1 + n_2\mathbf{a}_2 + n_3\mathbf{a}_3, \tag{5.31}$$

where $n_1$, $n_2$, and $n_3$ are integers. The corresponding reciprocal lattice vectors are defined by

$$\mathbf{G} = m_1\mathbf{b}_1 + m_2\mathbf{b}_2 + m_3\mathbf{b}_3, \tag{5.32}$$

where $m_1$, $m_2$, and $m_3$ are integers, and the fundamental basis vectors are

$$\left. \begin{aligned} \mathbf{b}_1 &= (2\pi/\tau)\mathbf{a}_2 \times \mathbf{a}_3 \\ \mathbf{b}_2 &= (2\pi/\tau)\mathbf{a}_3 \times \mathbf{a}_1 \\ \mathbf{b}_3 &= (2\pi/\tau)\mathbf{a}_1 \times \mathbf{a}_2 \end{aligned} \right\} \tag{5.33}$$

with $\tau = |\mathbf{a}_1 \cdot (\mathbf{a}_2 \times \mathbf{a}_3)|$ being the volume of the primitive unit cell defined by

$\mathbf{a}_1$, $\mathbf{a}_2$, and $\mathbf{a}_3$. It is apparent from their definition, eqn (5.33), that

$$\mathbf{a}_i \cdot \mathbf{b}_j = 2\pi\delta_{ij}, \qquad (5.34)$$

where $\delta_{ij} = 1$ for i = j but zero otherwise.

The phase factor in eqn (5.30) only defines the Bloch vector within a reciprocal lattice vector $\mathbf{G}$, since it follows from eqs (5.31)–(5.34) that $\mathbf{G} \cdot \mathbf{R}$ is an integer multiple of $2\pi$. Just as in the one-dimensional case it is customary to label the wave function $\psi_k$ by restricting $\mathbf{k}$ to lie within the first Brillouin zone, which is the closed volume about the origin in reciprocal space formed by bisecting near-neighbour reciprocal lattice vectors. For example, consider the simple cubic lattice with basis vectors $\mathbf{a}_1$, $\mathbf{a}_2$, $\mathbf{a}_3$ along the Cartesian axes x, y, z respectively. Because $a_1 = a_2 = a_3 = a$, it follows from eqn (5.33) that the reciprocal space basis vectors, $\mathbf{b}_1$, $\mathbf{b}_2$, $\mathbf{b}_3$, also lie along x, y, z respectively but with magnitude $2\pi/a$.

Thus, the reciprocal lattice of a simple cubic lattice is also simple cubic. It is shown in Fig. 5.7 in the xy plane, where it is clear that the bisectors of the first nearest-neighbour (100) reciprocal lattice vectors from a closed volume about the origin which is not cut by the second or any further near-neighbour bisectors. Hence, the Brillouin zone is a cube of volume $(2\pi/a)^3$ that from eqn (2.38) contains as many allowed $\mathbf{k}$ points as there are primitive unit cells in the crystal. The second, third, and fourth zones can

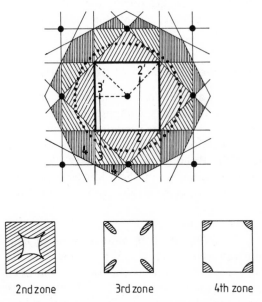

2nd zone          3rd zone          4th zone

**Fig. 5.7** The first four zones of the simple cubic lattice corresponding to $k_z = 0$. The dotted circle represents the cross-section of a spherical Fermi surface.

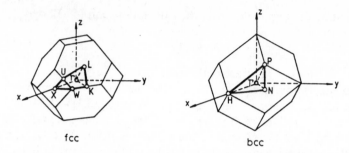

fcc                                    bcc

**Fig. 5.8** The fcc and bcc Brillouin zones. The symbol $\Gamma$ labels the centre of the zone. The intersections of the $|100\rangle$ and $|111\rangle$ directions with the Brillouin zone boundary are labelled $X$ and $L$ in the fcc case and $H$ and $P$ in the bcc case.

be 'folded-back' into the first zone through the action of appropriate reciprocal lattice vectors as illustrated in Fig. 5.7. Consequently, as shown in the lower part of Fig. 5.7, the spherical Fermi surface of a free-electron gas would be broken up into segments by the presence of a weakly perturbing periodic potential that opens up small energy gaps across the zone boundaries.

The Brillouin zones for the fcc and bcc lattices are drawn in Fig. 5.8. We see that it is customary to assign high symmetry **k** points with specific symbols. Thus, the centre of the Brillouin zone is labelled $\Gamma$, whereas the intersections of the $|100\rangle$ and $|111\rangle$ directions with the zone boundary are labelled $X$ and $L$ in the fcc case, $H$ and $P$ in the bcc case respectively.

## 5.4. The nearly free electron approximation

The bandstructure of fcc aluminium is shown in Fig. 5.9 along the directions $\Gamma X$ and $\Gamma L$ respectively. It was computed by solving the Schrödinger equation selfconsistently within the local density approximation (LDA). We see that aluminium is indeed a NFE metal in that only small energy gaps have opened up at the Brillouin zone boundary. We may, therefore, look for an approximate solution to the Schrödinger equation that comprises the linear combination of only a *few* plane waves, the so-called NFE approximation.

In particular, let us consider the band structure along $\Gamma X$ where $\mathbf{k}_\Gamma = (0, 0, 0)$ and $\mathbf{k}_X = (2\pi/a)(1, 0, 0)$ with $a$ the edge length of the face-central cubic unit cell. (Note that the $X$ point for fcc is $2\pi/a$ not $\pi/a$ like for simple cubic.) In this direction the two lowest free-electron bands correspond to $E_{\mathbf{k}} = (\hbar^2/2m)\mathbf{k}^2$ and $E_{\mathbf{k}+\mathbf{g}} = (\hbar^2/2m)(\mathbf{k} + \mathbf{g})^2$ respectively. The term $\mathbf{g}$ is the reciprocal lattice vector $(2\pi/a)(\bar{2}, 0, 0)$ that 'folds-back' the free-electron states into the Brillouin zone along $\Gamma X$, so that $E_{\mathbf{k}}$ and $E_{\mathbf{k}+\mathbf{g}}$

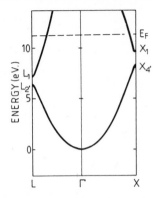

**Fig. 5.9** The band structure of fcc aluminium. (After Moruzzi *et al.* (1978).)

are degenerate at $X$ (cf $|k_X + g| = |k_X|$). We expect the weak underlying potential of the fcc aluminium lattice to mix those two free-electron states together. Hence, we look for the NFE solution

$$\psi_k = c_1 \psi_k^{(1)} + c_2 \psi_k^{(2)}, \tag{5.35}$$

where

$$\psi_k^{(1)} = L^{-3/2} e^{ik \cdot r} \tag{5.36}$$

and

$$\psi_k^{(2)} = L^{-3/2} e^{i(k+g) \cdot r}. \tag{5.37}$$

Substituting eqn (5.35) into the Schrödinger equation, premultiplying by $\psi_k^{(1)*}$ or $\psi_k^{(2)*}$, and integrating over the volume of the crystal yields the NFE secular equation

$$\begin{bmatrix} \dfrac{\hbar^2}{2m} k^2 - E & v(200) \\[2ex] v(200) & \dfrac{\hbar^2}{2m} (k+g)^2 - E \end{bmatrix} \begin{bmatrix} c_1 \\[2ex] c_2 \end{bmatrix} = 0. \tag{5.38}$$

The element $v(200)$ is the $(2\pi/a)(2, 0, 0)$ Fourier component of the crystalline potential normalized by the volume of the crystal, namely

$$v(q) = L^{-3} \int V(r) e^{-iq \cdot r} \, dr. \tag{5.39}$$

The element $v(200)$ is real due to the symmetry of the fcc lattice. The energy, $E$, in eqn (5.38) is measured with respect to the average potential, $v(000)$.

Non-trivial solutions to the NFE secular equation exist if

$$
\begin{vmatrix}
\dfrac{\hbar^2}{2m}\,k^2 - E & v(200) \\[2ex]
v(200) & \dfrac{\hbar^2}{2m}\,(\mathbf{k} + \mathbf{g})^2 - E
\end{vmatrix} = 0. \tag{5.40}
$$

This quadratic equation has solutions

$$
E_{\mathbf{k}} = \frac{1}{2}\left[\frac{\hbar^2}{2m}\,k^2 + \frac{\hbar^2}{2m}\,(\mathbf{k} + \mathbf{g})^2\right]
$$

$$
\pm\, \frac{1}{2}\left\{\left[\frac{\hbar^2}{2m}\,(\mathbf{k} + \mathbf{g})^2 - \frac{\hbar^2}{2m}\,k^2\right]^2 + 4[v(200)]^2\right\}^{1/2}. \tag{5.41}
$$

Therefore, at the zone boundary $X$ where $k^2 = (\mathbf{k} + \mathbf{g})^2$, the eigenvalues are given by

$$
E_X^{\pm} = (\hbar^2/2m)(2\pi/a)^2 \pm v(200), \tag{5.42}
$$

and the eigenfunctions are given from eqs (5.35) and (5.38) by

$$
\psi_X^{\pm} = \left(\frac{2}{L^3}\right)^{1/2}\begin{cases}(e^{i2\pi x/a} + e^{-i2\pi x/a})/2 \\ (e^{i2\pi x/a} - e^{-i2\pi x/a})/2i\end{cases} = \left(\frac{2}{L^3}\right)^{1/2}\begin{cases}\cos(2\pi x/a) \\ \sin(2\pi x/a).\end{cases} \tag{5.43}
$$

Thus, the presence of the periodic potential has opened up a gap in the free-electron band structure with energy separation

$$
E_{\text{gap}} = 2|v(200)|. \tag{5.44}
$$

From Fig. 5.9 the energy gap of aluminium at $X$ is about 1 eV, so that eqn (5.44) implies that the magnitude of the Fourier component of the potential is only 0.5 eV. This is small compared to the free-electron Fermi energy of more than 10 eV for aluminium. Hence the band structure $E_{\mathbf{k}}$ and the corresponding density of states, $n(E)$, are nearly-free-electron-like to a very good approximation.

The NFE behaviour has been observed experimentally in studies of the Fermi surface, the surface of constant energy, $E_F$, in $\mathbf{k}$ space which separates filled states from empty states at the absolute zero of temperature. It is found that the Fermi surface of aluminium is indeed very close to that of a spherical free-electron Fermi surface that has been folded back into the Brillouin zone in a manner not too dissimilar to that discussed earlier for the simple cubic lattice. Moreover, just as illustrated in Fig. 5.7 for the latter case, aluminium is found to have a large second-zone pocket of *holes* but smaller third- and fourth-zone pockets of electrons. This accounts very beautifully for the fact that aluminium has a *positive* Hall coefficient rather than the negative value expected for a gas of negatively charged free carriers (see, for example, Kittel (1986)).

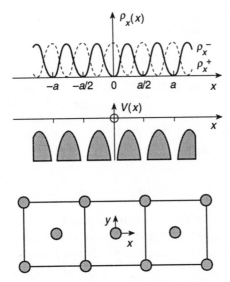

**Fig. 5.10** The upper figure shows the electron density corresponding to the eigenfunctions $\psi_X^+$ and $\psi_X^-$ at the fcc Brillouin zone boundary, $X$. We see that $\psi_X^+$ piles up charge into the core regions where the potential, $V(x)$, in the middle panel is large and negative. $V(x)$ is the averaged potential over all the atoms in the $yz$ plane of the fcc lattice shown in the lower figure.

## 5.5. Pseudopotentials

There is a major difficulty with the NFE picture which we have presented above for the sp-valent metals that is glossed over in most text books. The state at the bottom of the energy gap at $X$ in Fig. 5.9 is $X_{4'}$, which has p-like symmetry, whereas the state at the top of the energy gap is $X_1$, which has s-like symmetry (see, for example, Tinkham (1964)). We have seen from eqn (5.43) that the predicted NFE eigenfunctions are the two standing waves, $\cos k_X x$ and $\sin k_X x$, that result from the interference of the right- and left-travelling plane waves $e^{ik_X x}$ and $e^{-ik_X x}$ respectively. Their corresponding probability densities, $\rho_X^+ = |\psi_X^+|^2$ and $\rho_X^- = |\psi_X^-|^2$, are plotted in the upper part of Fig. 5.10. We see that $\rho_X^+$ piles up charge onto the atomic centres in a manner that is characteristic of s orbitals (cf $R_{ns}$ $(r = 0) \neq 0$ in Fig. 2.12). On the other hand, $\rho_X^-$ has zero charge at the atomic centres in a manner that is characteristic of p orbitals (cf $R_{2p}$ $(r = 0) = 0$ in Fig. 2.12).

Thus, the p-like state, $X_{4'}$, at the bottom of the energy gap must be associated with the NFE eigenfunction, $\rho_X^-$, which pushes charge away from the atomic centres, whereas the s-like state, $X_1$, at the top of the energy gap must be associated with the NFE eigenfunction, $\rho_X^+$, which pulls charge onto the atomic centres. From eqn (5.42) it follows that in order for the NFE approximation to fit the observed band structure, the (200) Fourier component of the potential must be repulsive. In fact, for aluminium with a band gap of 1 eV we have

$$v^{\text{NFE}}(200) = +0.5 \, \text{eV}. \tag{5.45}$$

But this contradicts our picture of the crystalline potential, $V(x)$, that is sketched in the middle of Fig. 5.10. It comprises to a good approximation the overlapping of attractive atomic potentials like those drawn for the diatomic molecule in Fig. 3.1. The (200) Fourier component of such a potential is large and negative, the *ab initio* crystalline potential for aluminium taking the value

$$v(200) \approx -5 \, \text{eV}. \tag{5.46}$$

Thus, we have a paradox: although the band structure of aluminium is nearly-free-electron-like, the actual Fourier components of the crystalline potential are large and negative, not small and positive as required by the NFE model.

The resolution of this paradox is easily obtained once it is remembered that the NFE bands in aluminium are formed from the *valence* 3s and 3p electrons. These states must be orthogonal to the s and p *core* functions, so that they contain nodes in the core region as illustrated for the 2s wave function in Fig. 2.12. In order to reproduce these very short wavelength oscillations, plane waves of very high momentum must be included in the plane wave expansion of $\psi_k$. Retaining only the two lowest energy plane waves in eqn (5.35) provides an extremely bad approximation.

In 1940 Herring circumvented this problem by starting at the outset with a basis of plane waves that had *already* been orthogonalized to the core states, the so-called orthogonalized plane wave (OPW) basis. Retaining only the two lowest orthogonalized plane waves we can look for the OPW solution that is analogous to eqn (5.35), namely

$$\psi_k = c_1 \chi_k^{(1)} + c_2 \chi_k^{(2)}, \tag{5.47}$$

where $\chi_k^{(1)}$ and $\chi_k^{(2)}$ are the orthogonalized plane waves

$$\chi_k^{(\alpha)} = \psi_k^{(\alpha)} - \sum_c \beta_c^{(\alpha)} \psi_c \quad (\alpha = 1, 2). \tag{5.48}$$

The sum $c$ runs over all the core states $\psi_c$ within the crystal. The prefactor $\beta_c^{(\alpha)}$ is chosen to guarantee that $\chi_k^{(\alpha)}$ is indeed orthogonal to any core state, that is

$$\int \psi_c^* \chi_k^{(\alpha)} \, d\mathbf{r} = 0. \tag{5.49}$$

Substituting eqn (5.48) into eqn (5.49) and using the fact that the core states are all orthogonal to each other we have at once

$$\beta_c^{(\alpha)} = \int \psi_c^* \chi_{\mathbf{k}}^{(\alpha)} \, d\mathbf{r} = S_{c\mathbf{k}_\alpha}. \tag{5.50}$$

That is, the prefactor, $\beta_c^{(\alpha)}$, is the overlap integral between the core state $\psi_c$ and the plane wave state with wave vector $\mathbf{k}_\alpha$ (where $\mathbf{k}_1 = \mathbf{k}$ and $\mathbf{k}_2 = (\mathbf{k} + \mathbf{g})$).

The energy gap arises from the off-diagonal element in the resulting $2 \times 2$ OPW secular equation coupling the states $\chi_{\mathbf{k}}^{(1)}$ and $\chi_{\mathbf{k}}^{(2)}$ together. From eqn (5.48) it has the form

$$\int \chi_{\mathbf{k}}^{(1)*}(\hat{H} - E)\chi_{\mathbf{k}}^{(2)} \, d\mathbf{r}$$

$$= \left\{ \begin{array}{l} \int \psi_{\mathbf{k}}^{(1)*}(\hat{H} - E)\psi_{\mathbf{k}}^{(2)} \, d\mathbf{r} - \sum_{c'} \beta_{c'}^{(2)} \int \psi_{\mathbf{k}}^{(1)*}(\hat{H} - E)\psi_{c'} \, d\mathbf{r} \\[2mm] - \sum_c \beta_c^{(1)*} \int \psi_c^*(\hat{H} - E)\psi_{\mathbf{k}}^{(2)} \, d\mathbf{r} + \sum_{c,c'} \beta_c^{(1)*}\beta_{c'}^{(2)} \int \psi_c^*(\hat{H} - E)\psi_{c'} \, d\mathbf{r} \end{array} \right\}$$

$$\tag{5.51}$$

The first term on the right-hand side is just the usual NFE off-diagonal element, namely $v(200)$ (cf eqn (5.38)). The remaining three terms on the right-hand side may be grouped together by using the Schrödinger equation for the core states, namely

$$(\hat{H} - E)\psi_c = E_c \psi_c, \tag{5.52}$$

where $E_c$ is the energy level of the core state $c$. Remembering that $\hat{H}$ is a hermitian operator that may act to either the right or to the left, eqn (5.51) simplifies to

$$\int \chi_{\mathbf{k}}^{(1)*}(\hat{H} - E)\chi_{\mathbf{k}}^{(2)} \, d\mathbf{r} = v(200) + \sum_c (E - E_c)\beta_c^{(1)*}\beta_c^{(2)}. \tag{5.53}$$

The energy gap at $X$ will be given by twice the modulus of this off-diagonal matrix element for $\mathbf{k} = \mathbf{k}_X$. Thus, within the OPW approximation we can write

$$E_{\text{gap}} = 2\left| v(200) + \sum_c (E_X - E_c)S_{\mathbf{k}_Xc}^2 \right|, \tag{5.54}$$

where we have replaced $E$ on the right-hand side of eqn (5.53) by the energy at the centre of the gap, namely, $E_X$, which is an excellent approximation for small gaps. Since $E_X - E_c > 0$, the core-orthogonality term in eqn (5.54) is repulsive, the individual core contributions being proportional to the square of their overlap integrals (choosing the core functions $\psi_c$ to be real).

It is yet again another manifestation of Pauli's exclusion principle, valence electrons being forbidden from entering core states that are already occupied.

This resolves the paradox. The energy gap at $X$ is small because the large, negative Fourier component of the crystalline potential, $v(200)$, is countered by an equally large but positive core-orthogonality contribution. Thus, we can retain the NFE description of the sp-valent metals, provided we replace the true crystalline potential, $V(\mathbf{r})$, by a *pseudopotential*, $V_{ps}(\mathbf{r})$, which includes intrinsically the core-orthogonality effects. In practice there is almost total cancellation between the repulsive core-orthogonality term and the attractive coulomb potential within the core region. We will, therefore, approximate the ionic pseudopotential using the Ashcroft empty core pseudopotential, namely

$$v_{ps}^{ion}(\mathbf{r}) = \begin{cases} 0 & \text{for } r < R_c \\ -Ze^2/4\pi\varepsilon_0 r & \text{for } r > R_c, \end{cases} \tag{5.55}$$

where $Z$ is the valence. This is drawn in Fig. 5.11.

The Fourier components of the resulting ionic lattice will oscillate in sign, since from eqn (5.39)

$$v_{ps}^{ion}(\mathbf{q}) = -(Ze^2/\varepsilon_0\Omega)(\cos qR_c)/q^2 \tag{5.56}$$

where $\Omega$ is the volume per atom. In the absence of the core the Fourier components are negative as expected, but in the presence of the core the Fourier components may become positive. As $q$ increases from zero, the first change from negative to positive behaviour occurs at $q_0$, such that $\cos q_0R_c = 0$, that is

$$q_0 = \frac{\pi}{2R_c}. \tag{5.57}$$

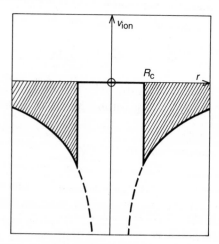

**Fig. 5.11** The Ashcroft empty core pseudopotential.

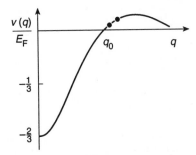

**Fig. 5.12** The Heine–Abarenkov (1964) pseudopotential for aluminium which has been normalized by the Fermi energy. The term $q_0$ gives the position of the first node. The two large dots mark the zone boundary Fourier components, $v_{ps}(111)$ and $v_{ps}(200)$.

As is seen from the behaviour of the more sophisticated Heine–Abarenkov pseudopotential in Fig. 5.12, the first node $q_0$ in aluminium lies just to the left of $(2\pi/a)\sqrt{3}$ and $g = (2\pi/a)2$, the magnitude of the reciprocal lattice vectors that determine the band gaps at $L$ and $X$ respectively. This explains both the positive value and the smallness of the Fourier component of the potential, which we deduced from the observed band gap in eqn (5.45). Taking the equilibrium lattice constant of aluminium to be $a = 7.7$ au and reading off from Fig. 5.12 that $q_0 \simeq 0.8(4\pi/a)$, we find from eqn (5.57) that the Ashcroft empty core radius for aluminium is $R_c = 1.2$ au. Thus, the ion core occupies only 6% of the bulk atomic volume. Nevertheless, we will find that its strong repulsive influence has a marked effect not only on the equilibrium bond length but also on the crystal structure adopted.

Figure 5.13 shows the densities of states $n(E)$ for the sp-valent metals that have been computed by solving the Schrödinger equation self-consistently within the local density approximation. We see that Na, Mg, and Al across the second period and Al, Ga, and In down group III are good NFE metals because their densities of states are only very small perturbations of the free-electron densities of states shown in Fig. 2.9(b). However, we see that Li and Be display very strong deviations from free-electron behaviour. This is a direct consequence of these first-row elements having no p core electrons, so that from eqn (5.54) there is no repulsive core-orthogonality component to cancel the attractive coulomb potential which 2p-like states such as $X_{4'}$ feel. This leads to sizeable Fourier components of the pseudopotential and hence very large band gaps, the p-like state at the bottom of the gap being lowered considerably compared to its free-electron value. For example, in fcc Be the gap at $L$ is 5.6 eV compared to the Al gap of only 0.34 eV in Fig. 5.9. In fact, the band gaps in different directions at the Brillouin zone boundary are nearly large enough for a gap to open up in the Be density of states at the Fermi energy in Fig. 5.13, thereby leading to semiconducting

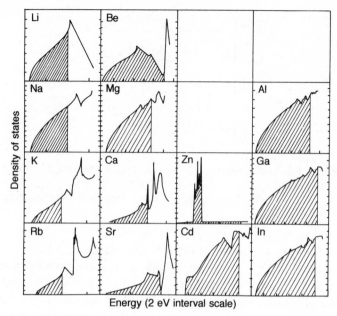

**Fig. 5.13** The densities of states of sp-valent metals. (After Moruzzi *et al.* (1978).)

behaviour. It is this absence of core p electrons and their corresponding repulsive influence that causes the interatomic potentials for the first-row elements such as C, N, and O to be characterized by softer normalized cores, as discussed earlier in Chapter 3.

The occupied energy levels of the heavier alkalis K and Rb and the alkaline earths Ca and Sr are affected by the presence of the respective 3d and 4d band, which lies just above the Fermi energy (cf the relative positions of the s and d free atom energy levels in Fig. 2.17). In particular, we see from Fig. 5.13 that a gap has nearly opened up at the Fermi energy in Sr. Theoretically it is predicted that Sr becomes a semiconductor at 0.3 GPa of pressure, which agrees reasonably with high-pressure resistivity data (Jan and Skriver (1981)). The group IIB elements Zn and Cd, on the other hand, have their valence states strongly distorted by the presence of the filled d band. In Fig. 2.17 we see that the 5s–4d energy separation in Cd is larger than the 4s–3d separation in Zn, which results in the Cd 4d band lying about 1 eV below the bottom of the valence 5sp band. We will see in the next chapter that the proper inclusion of d states within the pseudopotential is required to explain why Ca and Sr are fcc but Zn and Cd are hcp with their large axial ratios of 1.86 and 1.89 respectively. Such an *l*-dependent pseudo-potential is said to be *non-local*, whereas the simple Ashcroft empty-core pseudopotential, which does not differentiate explicitly between states of different angular momentum, is said to be *local* (see, for example, Ashcroft and Mermin (1976)).

## 5.6. The nature of the metallic bond in sp-valent metals

Wigner and Seitz were the first to apply quantum mechanics to a study of the metallic bond in sodium in 1933. They began by locating the bottom of the conduction band which is determined by the average potential, $v(000)$. They argued that since the bottom of the band corresponded to the most bonding state, it satisfied the bonding boundary condition that the gradient of the wave function across the boundary of the Wigner–Seitz cell vanished. This cell is formed in real space about a given atom by bisecting the near-neighbour position vectors in the same way that the Brillouin zone is formed in reciprocal space. The Wigner–Seitz cell of the bcc lattice is the fcc Brillouin zone and vice versa (see, for example, Kittel (1986)). Since there are twelve nearest neighbours in the fcc lattice and fourteen first and second nearest neighbours in the bcc lattice, it is a very good approximation to replace the Wigner–Seitz cell by a Wigner–Seitz sphere of the same volume (cf Fig. 5.8).

The energy of the bottom of the sodium conduction band, denoted by $\Gamma_1$, is determined by imposing the bonding boundary condition across the Wigner–Seitz sphere of radius, $R_{WS}$, namely

$$[dR_{3s}(r, E)/dr]_{r=R_{WS}, E=\Gamma_1} = 0 \qquad (5.58)$$

where $R_{3s}(r, E)$ is the $n = 3, l = 0$ solution of the radial Schrödinger equation within the Wigner–Seitz sphere. The free sodium atom radial function, $R_{3s}(r)$, is the solution of the same Schrödinger equation but with the different boundary condition that $R_{3s}(r) \rightarrow 0$ as $r \rightarrow \infty$. Having located the bottom of the conduction band, Wigner and Seitz then added to it the average kinetic energy per electron of a free-electron gas, in order to obtain the total binding energy per atom as shown in Fig. 5.14. They found values of the cohesive energy, equilibrium atomic volume, and bulk modulus that were within 10% of experiment.

We can understand the behaviour of the binding energy curves of monovalent sodium and other polyvalent metals by considering the metallic bond as arising from the immersion of an ionic lattice of empty core pseudopotentials into a free-electron gas as illustrated schematically in Fig. 5.15. We have seen that the pseudopotentials will only perturb the free-electron gas weakly so that, as a first approximation, we may assume that the free-electron gas remains uniformly distributed throughout the metal. Thus, the total binding energy per atom may be written as

$$U = ZU_{eg} + U_{es}, \qquad (5.59)$$

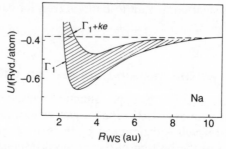

**Fig. 5.14** The binding energy $U$ as a function of the Wigner–Seitz radius $R_{WS}$ for sodium. The bottom of the conduction band, $\Gamma_1$, is given by the lower curve to which is added the average kinetic energy per electron (the shaded region). (After Wigner and Seitz (1933).)

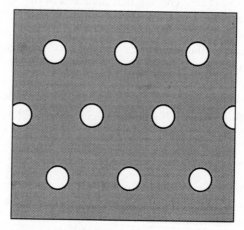

**Fig. 5.15** An ionic lattice of Ashcroft empty core pseudopotentials immersed in a free-electron gas.

where $Z$ is the valence. The quantity, $U_{eg}$, is the sum of the kinetic and exchange-correlation energies per electron that has already been given explicitly as a function of the average electron radius, $r_s$, through eqns (2.46), (2.47), and (2.48), namely

$$U_{eg} = \frac{2.210}{r_s^2} - \frac{0.916}{r_s} - (0.115 - 0.0313 \ln r_s), \qquad (5.60)$$

where atomic units have been used. $U_{es}$ is the electrostatic energy of interaction between the ions and the electrons. This is evaluated by neglecting the coulomb interaction between different Wigner–Seitz cells as they are electrically neutral and by approximating the electrostatic energy of a single

Wigner–Seitz cell by that of a sphere. Thus,

$$U_{es} = -\int \rho(\mathbf{r}) v_{ps}^{ion}(\mathbf{r})\, d\mathbf{r} + \frac{1}{2}\frac{e^2}{4\pi\varepsilon_0}\iint \frac{\rho(\mathbf{r})\rho(\mathbf{r}')}{|\mathbf{r}-\mathbf{r}'|}\, d\mathbf{r}\, d\mathbf{r}', \qquad (5.61)$$

where the integrals are carried out within the Wigner–Seitz sphere. The first contribution is the electron-ion interaction whereas the second contribution is the electron–electron interaction. Substituting into eqn (5.61) the Ashcroft empty core pseudopotential $v_{ps}^{ion}$ from eqn (5.55) and the density of the free-electron gas, $\rho = Z/\Omega = Z/(\tfrac{4}{3}\pi R_{WS}^3)$, we find

$$U_{es} = -\frac{3Z^2}{R_{WS}}\left[1-\left(\frac{R_c}{R_{WS}}\right)^2\right] + \frac{1.2Z^2}{R_{WS}}, \qquad (5.62)$$

since $e^2 = 2$ and $4\pi\varepsilon_0 = 1$ in atomic units.

Consider first the monovalent alkali metals when $Z = 1$ and $r_s = R_{WS}$. It then follows from eqs (5.60) and (5.62) that

$$U \approx -\frac{3}{R_{WS}}\left[1-\left(\frac{R_c}{R_{WS}}\right)^2\right] + \frac{2.210}{r_s^2}, \qquad (5.63)$$

since the electron–electron self-energy contribution $1.2/R_{WS}$ is almost exactly cancelled by the exchange-correlation contribution

$$-0.916/r_s - (0.115 - 0.0313 \ln r_s).$$

We see that eqn (5.63) mirrors the behaviour found in Fig. 5.14 for sodium by Wigner and Seitz. At metallic densities the bottom of the conduction band is well described by the first contribution in eqn (5.63). As the atoms are brought together, the bonding state $\Gamma_1$ becomes more bonding until eventually the repulsive core contribution dominates, and the bottom of the conduction band rises rapidly. From eqn (5.63) the maximum binding energy of this state $\Gamma_1$ occurs for

$$(R_{WS})_m = \sqrt{3}R_c. \qquad (5.64)$$

Since for sodium $R_c \simeq 1.7$ au, eqn (5.64) predicts that $\Gamma_1$ has a minimum at about 2.9 au. This is in good agreement with the curve in Fig. 5.14 that was obtained by solving the radial Schrödinger equation subject to the boundary condition eqn (5.58).

In general, the equilibrium Wigner–Seitz radius $R_{WS}^0$ can be found from eqn (5.59) by requiring that $U$ is stationary with respect to $R_{WS}$. It is found to depend explicitly on the core radius $R_c$ through the equation

$$\left(\frac{R_c}{R_{WS}^0}\right)^2 = \frac{1}{5} + \frac{0.102}{Z^{2/3}} + \frac{0.0035R_{WS}^0}{Z} - \frac{0.491}{Z^{1/3}R_{WS}^0}, \qquad (5.65)$$

where the first four terms are the coulomb, exchange, correlation, and kinetic contributions respectively. Girifalco (1976) has taken the experimental values of the Wigner–Seitz radius $R_{WS}^0$ to determine the Ashcroft empty core radius $R_c$ from eqn (5.65). The resultant values are listed in Table 5.2 where, as expected, the core size increases as we go down a given group in the periodic table but decreases as we go across a given period. It is clear from Table 5.2 that only sodium has an equilibrium value of $r_s$ that is close to the free-electron gas value of 4.2 au.

The equilibrium bulk modulus, which reflects the curvature of the binding energy curve through $B = V(\mathrm{d}^2U/\mathrm{d}V^2)$, may be written from eqs (5.59) and (5.65) in the form

$$B/B_{ke} = 0.200 + 0.815R_c^2/r_s, \qquad (5.66)$$

where the correlation contribution has been neglected since it contributes less

**Table 5.2** Equilibrium bulk properties of the simple and noble metals

| Metal | Z | $U_{coh}/Z$ (eV/electron) | $R_{WS}^0$ [a] (au) | $r_s$ [a] (au) | $R_c$ (au) | $B/B_{ke}$ (eqn 5.66) | $B/B_{ke}$ (expt.) |
|---|---|---|---|---|---|---|---|
| Li | 1 | 1.7 | 3.27 | 3.27 | 1.32 | 0.63 | 0.50 |
| Na | 1 | 1.1 | 3.99 | 3.99 | 1.75 | 0.83 | 0.80 |
| K  | 1 | 0.9 | 4.86 | 4.86 | 2.22 | 1.03 | 1.10 |
| Rb | 1 | 0.9 | 5.31 | 5.31 | 2.47 | 1.14 | 1.55 |
| Cs | 1 | 0.8 | 5.70 | 5.70 | 2.76 | 1.29 | 1.43 |
| Be | 2 | 1.7 | 2.36 | 1.87 | 0.76 | 0.45 | 0.27 |
| Mg | 2 | 0.8 | 3.35 | 2.66 | 1.31 | 0.73 | 0.54 |
| Ca | 2 | 0.9 | 4.12 | 3.27 | 1.73 | 0.95 | 0.66 |
| Sr | 2 | 0.9 | 4.49 | 3.57 | 1.93 | 1.05 | 0.78 |
| Ba | 2 | 0.9 | 4.67 | 3.71 | 2.03 | 1.11 | 0.84 |
| Zn | 2 | 0.7 | 2.91 | 2.31 | 1.07 | 0.60 | 0.45 |
| Cd | 2 | 0.6 | 3.26 | 2.59 | 1.27 | 0.71 | 0.63 |
| Hg | 2 | 0.3 | 3.35 | 2.66 | 1.31 | 0.73 | 0.59 |
| Al | 3 | 1.1 | 2.99 | 2.07 | 1.11 | 0.69 | 0.32 |
| Ga | 3 | 0.9 | 3.16 | 2.19 | 1.20 | 0.74 | 0.33 |
| In | 3 | 0.9 | 3.48 | 2.41 | 1.37 | 0.83 | 0.39 |
| Tl | 3 | 0.6 | 3.58 | 2.49 | 1.43 | 0.87 | 0.39 |
| Cu | 1 | 3.5 | 2.67 | 2.67 | 0.91 | 0.45 | 2.16 |
| Ag | 1 | 3.0 | 3.02 | 3.02 | 1.37 | 0.71 | 2.94 |
| Au | 1 | 3.8 | 3.01 | 3.01 | 1.35 | 0.69 | 4.96 |

[a] From Girifalco (1976).

than a few percent. The quantity, $B_{ke}$, is the bulk modulus of the non-interacting free-electron gas, namely

$$B_{ke} = 0.586/r_s^5. \tag{5.67}$$

It follows from eqn (5.66) and Table 5.2 that the presence of the ion core is crucial for obtaining realistic values of the bulk modulus of sp-valent metals. However, we see that the sd-valent noble metals Cu, Ag, and Au are not describable by the NFE approximation, the theoretical bulk modulus being a factor of five too small. We will return to the noble metals at the end of the next chapter.

## 5.7. Embedded atom potentials

This concept of metallic cohesion as arising from embedding ions in a gas of free electrons suggests that the binding energy of a collection of atoms with position vectors $\mathbf{R}_i$ may be approximated in the form of an *embedded atom potential*, namely

$$U = \tfrac{1}{2}\sum_{i,j}' \Phi(R_{ij}) + \sum_i F(\rho_i) \tag{5.68}$$

where

$$\rho_i = \sum_j' \rho_{atom}(R_{ij}) \tag{5.69}$$

(Daw and Baskes 1984; Finnis and Sinclair 1984; Jacobsen *et al.* 1987). The first term $\Phi(R)$ is a *pairwise* interatomic potential that represents the electrostatic interaction and overlap repulsion between neighbouring atoms. The second term $F(\rho_i)$ is a *many-body* embedding potential that represents the energy of embedding an atom at site $i$ in the local charge density $\rho_i$ which comes from the tails of the atomic charge clouds $\rho_{atom}$ on neighbouring sites $j$.

The behaviour of the embedding function $F(\rho)$ is illustrated in Fig. 5.16 for the case of hydrogen and the rare-gas atoms helium and neon. The curves were evaluated within the local density approximation (LDA) by embedding the atom in a homogeneous electron gas of density $\rho$. We see that He and Ne display a positive embedding energy at all densities because their *closed* electronic shells repel the free electron gas through orthogonality constraints. Moreover, the variation with density is linear so that from eqn (5.69)

$$F_{closed}(\rho_i) = a\rho_i = a \sum_j' \rho_{atom}(R_{ij}) \tag{5.70}$$

where $a$ is an element dependent constant. Thus, the embedding function is *pairwise* and may be combined with the first contribution in eqn (5.68), leading to the well-known result that the interaction between rare-gas atoms is accurately described by pair potentials. On the other hand, the *open*-shell hydrogen atom shows a minimum in Fig. 5.16 at an attractive embedding

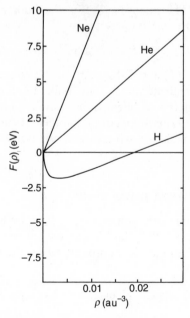

**Fig. 5.16** The embedding energy $F(\rho)$ for H and the rare gas atoms He and Ne in a free-electron gas of density $\rho$. (After Puska *et al.* (1981).)

energy of $-1.8\,\text{eV}$. The shape of this embedding curve is well reproduced by adding an attractive square root dependent term to the repulsive linear term of the closed-shell atoms, namely

$$F_{\text{open}}(\rho_i) = a\rho_i - b\rho_i^{1/2} \tag{5.71}$$

where $a$ and $b$ are positive constants. The presence of this second term leads to the non-pairwise, *many-body* character of the embedding function in open-shell systems.

This non-pairwise behaviour is most easily demonstrated by considering the coordination number dependence of the binding energy. It follows from eqs (5.68), (5.69), and (5.70) that the binding energy *per atom* of a lattice with coordination number $\varkappa$ may be written in the form

$$U = A\varkappa - B\varkappa^{1/2} \tag{5.72}$$

where

$$A = \tfrac{1}{2}\Phi(R_0) + a\rho_{\text{atom}}(R_0) \tag{5.73}$$

and

$$B = b\rho_{\text{atom}}^{1/2}(R_0) \tag{5.74}$$

with $R_0$ being the nearest neighbour distance. Thus, if we plot the binding energy of different lattices with identical nearest neighbour distances $R_0$ as

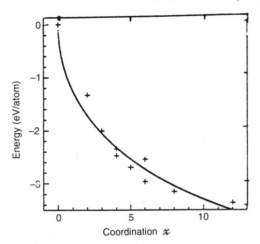

**Fig. 5.17** The binding energy per atom $U$ as a function of the coordination number $\varkappa$ for aluminium. The crosses correspond to LDA predictions, whereas the curve is a least-squares fit of the form of eqn (5.72). The lattice types considered are the linear chain ($\varkappa = 2$), graphite ($\varkappa = 3$), diamond ($\varkappa = 4$), two-dimensional square mesh ($\varkappa = 4$), square bilayer ($\varkappa = 5$), simple cubic ($\varkappa = 6$), triangular mesh ($\varkappa = 6$), vacancy lattice ($\varkappa = 8$) and face centred cubic ($\varkappa = 12$). (After Heine *et al.* (1991).)

a function of coordination number $\varkappa$ then we expect the results to fall on the curve given by eqn (5.72) since $A$ and $B$ will be constants. This is indeed the case for sp-valent aluminium where we see in Fig. 5.17 that the predicted LDA values of the binding energy of different lattices with nearest neighbour distances equal to that of fcc follow the simple curve of eqn (5.72) surprisingly well.

The square-root dependence of the attractive contribution to the binding energy is a consequence of the *unsaturated* nature of the metallic bond. If all the bonds were *saturated* then we would expect the binding energy to be directly proportional to the number of bonds present as the electrons in each bond would be *localized* between their parent atoms. However, in an sp-bonded metal there are not enough valence electrons to form saturated bonds with all close packed neighbours so that the electrons resonate between all bonds in a *delocalized* fashion. Increasing the local coordination about a given atom reduces the strength of neighbouring bonds as the electrons are spread more evenly between them. This is reflected in the fact that from eqn (5.72) the attractive binding energy *per bond* decreases as $1/\varkappa^{1/2}$ as the coordination number $\varkappa$ increases.

This has important consequences for the energetics of defects in metals. For example, the vacancy formation energy within a nearest neighbour *pair potential* approximation is given by

$$\Delta U_{\text{vac}}^{\text{pair}} = -z\Phi(R_0) + \tfrac{1}{2}z\Phi(R_0) \tag{5.75}$$

where the first term represents the loss in binding energy due to the breaking of $z$ bonds when the atom is removed to create the vacancy, whereas the second term represents the energy gain equal to the cohesive energy $U_{\text{coh}}$ when the removed atom is placed at the metal surface. We see, therefore, that a pair-potential model predicts a vacancy formation energy equal to the cohesive energy, i.e.

$$\Delta U_{\text{vac}}^{\text{pair}} = -\tfrac{1}{2}z\Phi(R_0) = U_{\text{coh}}. \tag{5.76}$$

This disagrees with experiment where the vacancy formation energy in metals is typically only about one-half the cohesive energy.

Using embedded atom potentials, however, there will be an additional lowering in the vacancy formation energy compared to that of pair potentials due to the fact that the $z$ atoms surrounding the vacancy have one less neighbour and, therefore, experience a bond strengthening with their remaining $(z-1)$ neighbours. This contributes an additional term

$$\Delta U_{\text{vac}}^{\text{add}} = z(z-1)\{-B[(z-1)^{-1/2} - z^{-1/2}]\}. \tag{5.77}$$

The prefactor outside the curly brackets gives the number of bonds that are strengthened by the absence of the atom at the vacancy site. The contribution inside the curly brackets gives the change in the bond energy due to the change in coordination from $z$ to $(z-1)$. For close-packed lattices $z \gg 1$ so that using the binomial expansion

$$\Delta U_{\text{vac}}^{\text{add}} \approx -\tfrac{1}{2}Bz^{1/2} = -\tfrac{1}{2}U_{\text{coh}} \tag{5.78}$$

where we have assumed that the cohesive energy is provided mainly by the attractive contribution $-Bz^{1/2}$ in the binding energy. Thus, the embedded atom potential predicts the total vacancy formation energy

$$\Delta U_{\text{vac}} \approx U_{\text{coh}} - \tfrac{1}{2}U_{\text{coh}} = \tfrac{1}{2}U_{\text{coh}}. \tag{5.79}$$

This is in good agreement with experiment.

Embedded atom potentials have been extensively used for performing atomistic simulations of point, line and planar defects in metals and alloys (e.g. Vitek and Srolovitz 1989). The pair potential $\Phi(R)$, atomic charge density $\rho_{\text{atom}}(\mathbf{r})$, and embedding function $F(\rho)$ are usually *fitted* to reproduce the known equilibrium atomic volume, elastic moduli, and ground state structure of the perfect defect-free lattice. However, the *prediction* of ground state structure, especially the competition between the common metallic structure types fcc, bcc, and hcp, requires a more careful treatment of the pair potential contribution $\Phi(R)$ than that provided by the semiempirical embedded atom potential. This is considered in the next chapter.

# References

Ashcroft, N. W. (1966). *Physics Letters* **23**, 48.
Ashcroft, N. W. and Mermin, N. D. (1976). *Solid state physics.* Holt, Rinehart and Winston, Philadelphia.
Cohen, M. L. (1987). *Microclusters* (eds S. Sugano, Y. Nishina, and S. Ohnishi), page 2. Springer, Berlin.
Daw, M. S. and Baskes, M. T. (1984). *Physical Review* **B29**, 6443.
Finnis, M. W. and Sinclair, J. E. (1984). *Philosophical Magazine* **A50**, 45.
Gasiorowicz, S. (1974). *Quantum physics.* Wiley, New York.
Girifalco, L. A. (1976). *Acta Metallurgica* **24**, 759.
Heine, V. and Abarenkov, I. (1964). *Philosophical Magazine* **9**, 451.
Heine, V., Robertson, I. J., and Payne, M. C. (1991). *Philosophical Transactions of the Royal Society (London)* **A334**, 393.
Herring, C. (1940). *Physical Review* **57**, 1169.
Jacobsen, K. W., Nørskov, J. K., and Puska, M. J. (1987). *Physical Review* **B35**, 7423.
Jan, J.-P. and Skriver, H. L. (1981). *Journal of Physics* **F11**, 805.
Kittel, C. (1986). *Introduction to solid state physics.* Wiley, New York.
Moruzzi, V. L., Janak, J. F., and Williams, A. R. (1978). *Calculated electronic properties of metals.* Pergamon, New York.
Puska, M. J., Nieminen, R. M., and Manninen, M. (1981). *Physical Review* **B24**, 3037.
Sutton, A. P. (1993). *Electronic structure of materials.* Oxford University Press.
Tinkham, M. (1964). *Group theory and quantum mechanics.* McGraw-Hill, New York.
Vitek, V. and Srolovitz, D. J. (eds) (1989). *Atomistic simulations of materials: beyond pair potentials.* Plenum, New York.
Wigner, E. P. and Seitz, F. (1933). *Physical Review* **43**, 804.

# 6

# Structure of sp-valent metals

## 6.1. Introduction

The description of the metallic bond in the previous chapter assumed that the free-electron gas remained homogeneously distributed as the pseudo-potential lattice of ion cores was immersed within it. In practice, of course, the free electrons will respond to screen the ion cores, thereby leading to a modulation of the electron density. This modulated electron density then acts in turn on the underlying ionic lattice to give rise to a structure-dependent contribution in the total energy. We will see in this chapter that this small structurally dependent contribution to the binding energy can be evaluated quantitatively using second-order perturbation theory within the nearly free electron (NFE) approximation.

Second-order perturbation theory provides a *reciprocal space* representation of this structural energy, since it mixes together all plane wave states that are linked by the reciprocal lattice vectors of the underlying perturbing lattice. It is, therefore, the appropriate representation for describing those few NFE systems such as the Hume–Rothery electron phases that are stabilized in **k**-space by the unperturbed free-electron Fermi sphere making contact with a zone boundary. More frequently, ground-state structural stability is determined by interaction between nearest-neighbour shells of atoms in *real space*. The beauty of second-order perturbation theory is that it also allows this structural energy to be written as a sum over oscillatory interatomic pair potentials. We will use this real space representation to provide an explanation for the structural trends within the sp-valent metals that are observed in Table 1.1.

## 6.2. Screening: the Thomas–Fermi approximation

Metals are characterized by an infinite static dielectric constant, so that any potential disturbance will be screened out by the electron response. This can be demonstrated most simply by considering a point ion of charge, $Ze$, immersed at the origin in a free-electron gas of charge density $-e\rho_0$. The presence of the ion will induce a new electron density, $\rho(\mathbf{r})$, which can be related to the new potential, $V(\mathbf{r})$, by Poisson's equation

$$\nabla^2 V(\mathbf{r}) = e^2[Z\delta(\mathbf{r}) + \rho_0 - \rho(\mathbf{r})]/\varepsilon_0. \tag{6.1}$$

The three contributions to the charge density on the right-hand side are the positive charge $Ze$ at the origin (represented by the delta function, $\delta(\mathbf{r})$), the compensating positive background charge, $e\rho_0$, of jellium, and the perturbed electronic charge density, $-e\rho(\mathbf{r})$. Note that strictly we should refer to $V(\mathbf{r})$ as the potential energy, but we will maintain our previous custom and call $V(\mathbf{r})$ the potential. In this chapter we will use atomic units throughout, so that $e^2 = 2$, $4\pi\varepsilon_0 = 1$, and $\hbar^2/2m = 1$ (cf. Note on the choice of units, page (xi)). Hence, Poisson's equation can be written

$$\nabla^2 V(\mathbf{r}) = 8\pi[Z\delta(\mathbf{r}) - \delta\rho(\mathbf{r})], \tag{6.2}$$

where $\delta\rho(\mathbf{r})$ is the induced change in electron density, namely

$$\delta\rho(\mathbf{r}) = \rho(\mathbf{r}) - \rho_0. \tag{6.3}$$

We will solve eqn (6.2) within the Thomas–Fermi approximation by linking the change in electron density, $\delta\rho(\mathbf{r})$, to the local potential, $V(\mathbf{r})$. At equilibrium the chemical potential or Fermi energy must be constant everywhere as illustrated in Fig. 6.1, so that

$$T(\mathbf{r}) + V(\mathbf{r}) = E_F^0, \tag{6.4}$$

where $T(\mathbf{r})$ and $V(\mathbf{r})$ are the values of the kinetic and potential energies respectively at the position $\mathbf{r}$. The term, $E_F^0$, is the chemical potential of the undisturbed uniform electron gas of density $\rho_0$, which from eqn (2.40) is

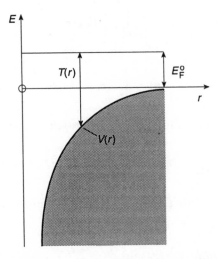

Fig. 6.1 The local potential energy, $V(r)$, and local kinetic energy, $T(r)$, such that the chemical potential is constant everywhere, taking the value $E_F^0$.

given by

$$E_F^0 = (3\pi^2 \rho_0)^{2/3}. \tag{6.5}$$

The Thomas–Fermi approximation assumes the variation in the potential, $V(\mathbf{r})$, to be sufficiently slow that the local kinetic energy, $T(\mathbf{r})$, is equal to that of an homogeneous free electron gas with the same density $\rho(\mathbf{r})$ as seen locally, that is

$$T(\mathbf{r}) = [3\pi^2 \rho(\mathbf{r})]^{2/3} = (1 + \delta\rho/\rho_0)^{2/3} E_F^0. \tag{6.6}$$

Hence, using the binomial expansion and substituting eqn (6.6) into (6.4), we have

$$E_F^0 [1 + \tfrac{2}{3}(\delta\rho/\rho_0) + \cdots] + V(\mathbf{r}) = E_F^0, \tag{6.7}$$

so that

$$\delta\rho(\mathbf{r}) = -(3\rho_0/2E_F^0)V(\mathbf{r}). \tag{6.8}$$

Thus, Poisson's equation (6.2) takes the simpler form

$$\nabla^2 V(\mathbf{r}) = 8\pi Z \delta(\mathbf{r}) + \kappa_{\mathrm{TF}}^2 V(\mathbf{r}), \tag{6.9}$$

where

$$\kappa_{\mathrm{TF}} = (12\pi\rho_0/E_F^0)^{1/2} = 2(3/\pi)^{1/6} \rho_0^{1/6}. \tag{6.10}$$

Poisson's equation (6.9) can be solved directly by writing the potential, $V(\mathbf{r})$, in terms of its Fourier transform, $V(\mathbf{q})$, that is

$$V(\mathbf{r}) = \frac{1}{(2\pi)^3} \int V(\mathbf{q}) \, e^{i\mathbf{q}\cdot\mathbf{r}} \, d\mathbf{q} \tag{6.11}$$

with

$$V(\mathbf{q}) = \int V(\mathbf{r}) \, e^{-i\mathbf{q}\cdot\mathbf{r}} \, d\mathbf{r}. \tag{6.12}$$

The delta function, $\delta(\mathbf{r})$, can similarly be written

$$\delta(\mathbf{r}) = \frac{1}{(2\pi)^3} \int e^{i\mathbf{q}\cdot\mathbf{r}} \, d\mathbf{q}, \tag{6.13}$$

since $\delta(\mathbf{q}) = 1$. Substituting eqs (6.11) and (6.13) into (6.9) we have

$$-q^2 V(\mathbf{q}) = 8\pi Z + \kappa_{\mathrm{TF}}^2 V(\mathbf{q}), \tag{6.14}$$

which has the solution

$$V(\mathbf{q}) = -8\pi Z/(q^2 + \kappa_{\mathrm{TF}}^2). \tag{6.15}$$

This has the well-known Fourier transform

$$V(\mathbf{r}) = -\frac{2Z}{r}\, e^{-\kappa_{TF} r}. \tag{6.16}$$

Thus, the ionic coulomb potential is damped exponentially within a Thomas–Fermi screening length $\lambda_{TF} = 1/\kappa_{TF}$. It follows from eqs (2.41) and (6.10) that

$$\lambda_{TF} = (\pi/12)^{1/3} r_s^{1/2}. \tag{6.17}$$

Screening in metals is very efficient: even the low-density metal sodium with $r_s = 4$ au has a Thomas–Fermi screening length as small as 1.3 au.

The screening in metals is also perfect. This follows from the fact that the total number of electrons associated with the screening density, $\delta\rho(\mathbf{r})$, is identically equal to $Z$, which can be seen by integrating

$$\delta\rho(\mathbf{r}) = Z(\kappa_{TF}^2/4\pi)\, e^{-\kappa_{TF} r}/r \tag{6.18}$$

over all space. The static dielectric constant, $\varepsilon(\mathbf{q})$, is defined through

$$V(\mathbf{q}) = V_{ext}(\mathbf{q})/\varepsilon(\mathbf{q}), \tag{6.19}$$

where $V_{ext}(\mathbf{q})$ is the external potential which the free-electron gas experiences, that is, $V_{ext}(\mathbf{q}) = -8\pi Z/q^2$ corresponds to the point ion potential, $V_{ext}(\mathbf{r}) = -2Z/r$. Substituting eqn (6.15) into (6.19) we see that the Thomas–Fermi dielectric constant is given by

$$\varepsilon_{TF}(\mathbf{q}) = 1 + \frac{\kappa_{TF}^2}{q^2}. \tag{6.20}$$

Thus the static dielectric constant diverges in the long wavelength limit as $q \to 0$.

## 6.3. Screening: linear response theory

The Thomas–Fermi approximation is, unfortunately, a poor approximation for the sp-valent metals. It is based on the assumption that the potential varies much more slowly than the screening length of the electrons themselves, so that the local approximation for the kinetic energy, eqn (6.6), is valid. In practice, however, the variation in the ionic potential is measured by the core radius, $R_c$ (cf Fig. 5.11), which is not large but of the same size as the screening length, $\lambda_{TF}$. Thus, we do not satisfy the criterion for the validity of the Thomas–Fermi approximation, since

$$R_c \gg \lambda_{TF}. \tag{6.21}$$

We will find that the Thomas–Fermi approximation totally fails to distinguish correctly between the different competing close-packed structure types such as fcc, bcc or hcp. We must, therefore, go beyond the Thomas–Fermi approximation and evaluate the proper screening behaviour of the free-electron gas at equilibrium metallic densities.

We have seen in the previous chapter that that the underlying ionic lattice only perturbs the electrons weakly. We may, therefore, use linear response theory to write

$$\delta\rho(\mathbf{q}) = \chi(\mathbf{q})V(\mathbf{q}) \tag{6.22}$$

that links the Fourier component of the screening electron density $\delta\rho(\mathbf{q})$ to the Fourier component of the total potential $V(\mathbf{q})$ through the linear response function $\chi(\mathbf{q})$. The dielectric constant $\varepsilon(\mathbf{q})$ may be expressed in terms of the linear response function $\chi(\mathbf{q})$ as follows. We have that

$$V(\mathbf{q}) = V_{ext}(\mathbf{q}) + \delta V(\mathbf{q}) = V_{ext}(\mathbf{q})/\varepsilon(\mathbf{q}) \tag{6.23}$$

where $\delta V(\mathbf{q})$ is the potential corresponding to the screening density $\delta\rho(\mathbf{q})$ induced by the presence of the external potential, $V_{ext}(\mathbf{q})$. The terms, $\delta V(\mathbf{q})$ and $\delta\rho(\mathbf{q})$, may be related to each other by taking the Fourier transform of Poisson's equation, namely

$$\delta V(\mathbf{q}) = \frac{8\pi}{q^2}\delta\rho(\mathbf{q}). \tag{6.24}$$

Substituting eqs (6.24) and (6.22) into eqn (6.23) we find that

$$V(\mathbf{q}) = V_{ext}(\mathbf{q})/[1 - (8\pi/q^2)\chi(\mathbf{q})], \tag{6.25}$$

so that finally from eqs (6.23) and (6.25) we have

$$\varepsilon(\mathbf{q}) = 1 - \frac{8\pi}{q^2}\chi(\mathbf{q}). \tag{6.26}$$

Within the Thomas–Fermi approximation, the linear response function is independent of the wavevector $\mathbf{q}$, since from eqn (6.20) it is given by

$$\chi_{TF}(\mathbf{q}) = -\kappa_{TF}^2/8\pi = \chi_{TF} \tag{6.27}$$

The wave vector dependence of the linear response function, $\chi(\mathbf{q})$, may be found by using perturbation theory to evaluate the change in the electronic density in the presence of the weakly perturbing potential

$$V(\mathbf{r}) = [V(\mathbf{q})\, e^{i\mathbf{q}\cdot\mathbf{r}} + \text{complex conjugate}], \tag{6.28}$$

where the complex conjugate is required to make the potential real. As proven in most textbooks on quantum mechanics (see, for example, Chapter 16 of Gasiorowicz (1974)) the addition of a perturbation $\hat{H}'$ to the Hamiltonian operator of a system causes the unperturbed eigenfunctions $\psi_k$ (with corresponding eigenvalues $E_k$) to mix together. To first order in the perturbation $\hat{H}'$ it is easily shown that the eigenfunctions of the perturbed system are given by

$$\psi_k^{(1)}(\mathbf{r}) = \psi_k(\mathbf{r}) + \sum_{k' \neq k} \frac{\hat{H}'_{k'k}}{E_k - E_{k'}} \psi_{k'}(\mathbf{r}), \tag{6.29}$$

where $\hat{H}'_{k'k}$ is the matrix element coupling the states, $\psi_k$ and $\psi_{k'}$, together through the perturbation, $\hat{H}'$, that is

$$\hat{H}'_{k'k} = \int \psi_{k'}^* \hat{H}' \psi_k \, d\mathbf{r}. \tag{6.30}$$

For the perturbing potential, $V(\mathbf{r})$, in eqn (6.28) this matrix element takes the form

$$\hat{H}'_{k'k} = V(\mathbf{q})\delta_{k',k+q} + V^*(\mathbf{q})\delta_{k',k-q}, \tag{6.31}$$

which follows immediately from eqn (6.30) on the imposition of periodic boundary conditions on the free-electron gas within a cube of side, $L$. Thus, substituting eqn (6.31) into eqs (6.29), we have

$$\psi_k^{(1)}(\mathbf{r}) = L^{-3/2}\left[ e^{i\mathbf{k}\cdot\mathbf{r}} + \frac{V(\mathbf{q})}{k^2 - (\mathbf{k}+\mathbf{q})^2} e^{i(\mathbf{k}+\mathbf{q})\cdot\mathbf{r}} + \frac{V^*(\mathbf{q})}{k^2 - (\mathbf{k}-\mathbf{q})^2} e^{i(\mathbf{k}-\mathbf{q})\cdot\mathbf{r}} \right]. \tag{6.32}$$

The change in the electron density is, therefore, given by

$$\delta\rho(\mathbf{r}) = \sum_k [|\psi_k^{(1)}(\mathbf{r})|^2 - |\psi_k(\mathbf{r})|^2] f_k, \tag{6.33}$$

where $f_k$ is the occupancy of the state $\mathbf{k}$. Hence, to first order in the perturbing potential,

$$\delta\rho(\mathbf{r}) = \left\{ L^{-3} \sum_k \left[ \frac{f_k}{k^2 - (\mathbf{k}+\mathbf{q})^2} + \frac{f_k}{k^2 - (\mathbf{k}-\mathbf{q})^2} \right] \right\}$$

$$\times [V(\mathbf{q}) e^{i\mathbf{q}\cdot\mathbf{r}} + \text{complex conjugate}]. \tag{6.34}$$

It follows from the definition of the linear response function in eqn (6.22) that

$$\chi_0(\mathbf{q}) = \frac{1}{L^3} \sum_k \frac{f_k - f_{k+q}}{k^2 - (\mathbf{k}+\mathbf{q})^2}, \tag{6.35}$$

where we have used the fact that $\mathbf{k}$ is a dummy variable that runs over all $\mathbf{k}$-space to replace $\mathbf{k} - \mathbf{q}$ by $\mathbf{k}'$ in the second summation in eqn (6.34).

## 142    Structure of sp-valent metals

The subscript zero on $\chi_0$ refers to the fact that we have performed our perturbation theory as though the electrons were independent particles. In practice, as we have seen in §2.5, the motion of each electron is correlated through the exchange-correlation hole. This leads to an enhancement of the response function which can be written

$$\chi(\mathbf{q}) = \chi_0(\mathbf{q})/[1 - I_{xc}\chi_0(\mathbf{q})] \tag{6.36}$$

where $I_{xc}$ is an enhancement factor that can be estimated within the local density approximation (see, for example, Taylor (1978)).

The summation in eqn (6.35) was first evaluated by Lindhard in 1954. Using the density of **k**-points given by eqn (2.38) the summation can be replaced by an integration

$$\frac{1}{L^3} \sum_{\mathbf{k}} \frac{f_{\mathbf{k}}}{k^2 - (\mathbf{k} + \mathbf{q})^2} = \frac{2}{(2\pi)^3} \int_0^{k_F} dk 2\pi k^2 \int_0^{\pi} \frac{\sin\theta\, d\theta}{-q^2 - 2kq\cos\theta}, \tag{6.37}$$

where the coordinate system used is shown in Fig. 6.2. This may be integrated directly to give

$$\frac{1}{L^3} \sum_{\mathbf{k}} \frac{f_{\mathbf{k}}}{k^2 - (\mathbf{k} + \mathbf{q})^2} = \frac{1}{4\pi^2 q} \int_0^{k_F} k \ln\left|\frac{q + 2k}{q - 2k}\right| dk$$

$$= -\frac{k_F}{4\pi^2}\left[\frac{1}{2} + \frac{1 - \eta^2}{4\eta} \ln\left|\frac{1 + \eta}{1 - \eta}\right|\right], \tag{6.38}$$

where $\eta = q/2k_F$. Replacing $\mathbf{q}$ by $-\mathbf{q}$ in the initial summation leaves the final result unchanged, so that from eqn (6.34) the Lindhard response function for the free-electron gas can be written

$$\chi_0(\eta = q/2k_F) = \left[\frac{1}{2} + \frac{1 - \eta^2}{4\eta} \ln\left|\frac{1 + \eta}{1 - \eta}\right|\right]\chi_{TF}, \tag{6.39}$$

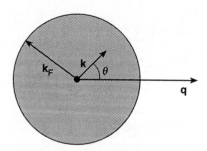

**Fig. 6.2** The coordinate system chosen for evaluating the linear response function, the integration being with respect to **k** whilst **q** is kept fixed.

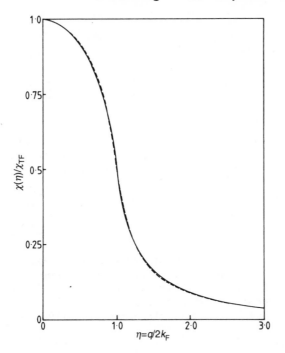

**Fig. 6.3** The wave-vector dependence of the Lindhard response function, $\chi(q/2k_F)$, which has been normalized by the constant Thomas–Fermi response function, $\chi_{TF}$. The dashed curve shows an approximation (eqn (6.89)) to the Lindhard response function that does not include the weak logarithmic singularity in the slope at $q/2k_F = 1$. (From Pettifor and Ward (1984).)

since the prefactor

$$-(k_F/2\pi^2) = -(\kappa_{TF}^2/8\pi) = \chi_{TF} \qquad (6.40)$$

from eqs (2.40), (6.17), and (7.27).

The wave vector dependence of the Lindhard function is shown in Fig. 6.3. We see that for very long wavelength disturbances as $q \to 0$ the response of the free-electron gas is that predicted by the Thomas–Fermi approximation as expected, since $2\pi/q \gg \lambda_{TF}$. However, we see that for very short wavelength disturbances as $q \to \infty$, the electrons are unable to screen out the potential, since the shortest unperturbed free-electron wavelength is $2\pi/k_F$. We observe that the response function has fallen to half of its Thomas–Fermi value at $q = 2k_F$. Although not discernible in Fig. 6.3, the slope of the curve diverges logarithmically for this value of $q = 2k_F$, as can be found directly by differentiating eqn (6.39). This very weak logarithmic singularity in **q**-space is picked up in the real-space Fourier transform of the electron density as very long range oscillations of wave vector $2k_F$. Thus,

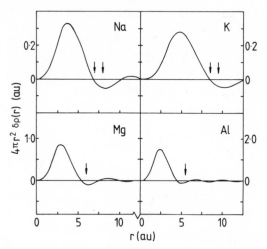

**Fig. 6.4** The radial charge distribution of the screening clouds around sodium, potassium, magnesium, and aluminium ions in free-electron environments of the appropriate equilibrium metallic densities. The arrows mark the positions of the first nearest neighbours in hcp Mg and fcc Al, the first and second nearest neighbours in bcc Na and K. (After Rasolt and Taylor (1975) and Dagens *et al.* (1975).)

rather than the screening cloud of electrons falling off exponentially around an ion core as in eqn (6.18), the screening cloud falls off algebraically as

$$\delta\rho(\mathbf{r}) \sim A_0 \cos 2k_{\mathrm{F}}r/r^3, \tag{6.41}$$

as first predicted by Friedel in 1952. Figure 6.4 shows the oscillations in the screening clouds around sodium, potassium, magnesium, and aluminium ions in free-electron environments of the appropriate equilibrium metallic densities. The coulomb interaction between these oscillatory screening clouds and the neighbouring ions gives rise to oscillatory interatomic pair potentials that account for the observed structural trends within the simple metals; we will examine these oscillatory interatomic pair potentials later in this chapter.

The physical origin of these *asymptotic* Friedel oscillations of wave vector, $2k_{\mathrm{F}}$, can be traced back to eqn (6.35) for the response function, $\chi_0(\mathbf{q})$. We see from the numerator that there are only contributions to the sum for the states, $\mathbf{k}$, that are occupied and the states $\mathbf{k} + \mathbf{q}$ that are unoccupied, or vice versa. This is to be expected considering Pauli's exclusion principle in that an electron in state, $\mathbf{k}$, can only scatter into state, $\mathbf{k} + \mathbf{q}$, if it is empty. Moreover, we see from the denominator in eqn (6.35) that the individual contributions will be largest for the case of scattering between states that are very close to the Fermi surface, since then $\mathbf{k}^2 - (\mathbf{k} + \mathbf{q})^2 \simeq 0$. We deduce from Fig. 6.5 that the maximum number of such scattering events will occur

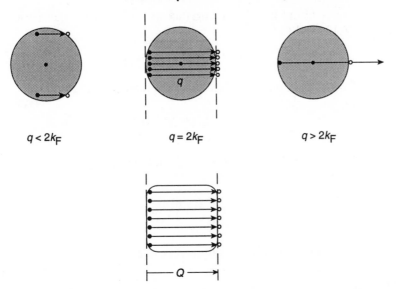

**Fig. 6.5** Scattering between filled and empty states near the Fermi surface. For the Fermi sphere of a free-electron gas the maximum number of such events occurs for $q = 2k_F$. For a Fermi surface with flat regions the number of such events is dramatically enhanced for $\mathbf{q} = \mathbf{Q}$, the spanning wave vector.

when $\mathbf{q}$ just spans the free-electron Fermi sphere, that is $q = 2k_F$. Although this leads to only a very weak logarithmic singularity in the slope of the linear response function of a free-electron gas, it can lead to much larger singularities in non-simple metal systems such as Cu–Pd alloys. In this latter case flat regions of the Fermi surface allow many states to contribute to the response function at the spanning wave vector, $\mathbf{Q}$. This leads to a so-called charge density wave of wave vector, $\mathbf{Q}$, that provides a modulation of the charge density, which is usually incommensurate with the underlying fcc lattice.

## 6.4. The reciprocal lattice representation

The total binding energy of a NFE metal can be evaluated within second-order perturbation theory. In the presence of a perturbation $\hat{H}'$ to the Hamiltonian operator of a system, the energy of state $\mathbf{k}$ is given by

$$E_k^{(2)} = E_k + \hat{H}'_{kk} + \sum_{k' \neq k} \frac{|\hat{H}'_{k'k}|^2}{E_k - E_{k'}} \qquad (6.42)$$

(see, for example, Chapter 16 of Gasiorowicz (1974)). The matrix element coupling states $\mathbf{k}$ and $\mathbf{k}' = \mathbf{k} + \mathbf{q}$ together in a NFE metal can be written from eqs (6.12) and (6.19) as

$$\hat{H}'_{\mathbf{k}+\mathbf{q},\mathbf{k}} = L^{-3} V(\mathbf{q}) = L^{-3} V^{\text{ion}}_{\text{ps}}(\mathbf{q})/\varepsilon(\mathbf{q}), \tag{6.43}$$

where $V^{\text{ion}}_{\text{ps}}(\mathbf{q})$ is the Fourier transform of the ionic potential

$$V^{\text{ion}}_{\text{ps}}(\mathbf{r}) = \sum_i v^{\text{ion}}_{\text{ps}}(\mathbf{r} - \mathbf{R}_i). \tag{6.44}$$

Substituting eqn (6.44) into (6.43) and interchanging the ordering of integration and summation, we have

$$\hat{H}'_{\mathbf{k}+\mathbf{q},\mathbf{k}} = [\varepsilon(\mathbf{q})]^{-1} \sum_i \frac{1}{\mathcal{N}} e^{-i\mathbf{q}\cdot\mathbf{R}_i} \left(\frac{\mathcal{N}}{L^3}\right) \int v^{\text{ion}}_{\text{ps}}(\mathbf{r} - \mathbf{R}_i) e^{-i\mathbf{q}\cdot(\mathbf{r}-\mathbf{R}_i)} d\mathbf{r}, \tag{6.45}$$

where we have introduced the total number of atoms, $\mathcal{N}$.

This can be expressed in a form which is familiar in X-ray diffraction, namely

$$\hat{H}'_{\mathbf{k}+\mathbf{q},\mathbf{k}} = S(\mathbf{q}) v_{\text{ps}}(\mathbf{q}), \tag{6.46}$$

where $S(\mathbf{q})$ is the *structure factor*

$$S(\mathbf{q}) = \mathcal{N}^{-1} \sum_i e^{-i\mathbf{q}\cdot\mathbf{R}_i}, \tag{6.47}$$

$v_{\text{ps}}(\mathbf{q})$ is the screened *pseudopotential form factor*,

$$v_{\text{ps}}(\mathbf{q}) = v^{\text{ion}}_{\text{ps}}(\mathbf{q})/\varepsilon(\mathbf{q}) = \left[\Omega^{-1} \int v^{\text{ion}}_{\text{ps}}(\mathbf{r}) e^{-i\mathbf{q}\cdot\mathbf{r}} d\mathbf{r}\right]/\varepsilon(\mathbf{q}), \tag{6.48}$$

and $\Omega$ is the volume per atom. For a periodic crystalline lattice with $n$ basis vectors $l_i$ $(i = 1, n)$ the structure factors simplifies to

$$S(\mathbf{q}) = \left[\frac{1}{n} \sum_i e^{-i\mathbf{G}\cdot l_i}\right] \delta_{q\mathbf{G}}, \tag{6.49}$$

that is, the structure factor vanishes unless $\mathbf{q} = \mathbf{G}$. For the Ashcroft empty-core pseudopotential the screened pseudopotential form factor can be written from eqs (5.56) and (6.20) as

$$v_{\text{ps}}(\mathbf{q}) = -\left(\frac{8\pi Z}{\Omega}\right) \frac{\cos qR_{\text{c}}}{q^2 + \kappa^2_{\text{TF}}}, \tag{6.50}$$

so that using eqn (2.40)

$$v_{\text{ps}}(\mathbf{q} = 0) = -\tfrac{2}{3} E^0_{\text{F}}. \tag{6.51}$$

Thus, the screened pseudopotential form factor of aluminium normalized by the Fermi energy will approach $q = 0$ at $-2/3$ as observed in Fig. 5.12.

The zeroth- and first-order contributions to the perturbed energy level, $E_k^{(2)}$, are independent of the particular arrangement of the atoms. The structure dependence resides in the second-order contribution. Summing over all the occupied energy levels the structure-dependent contribution takes the form

$$U_{bs}^{(2)} = \frac{1}{\mathcal{N}} \sum_k f_k \sum_q' \frac{|S(\mathbf{q})|^2 |v_{ps}(\mathbf{q})|^2}{k^2 - (\mathbf{k} + \mathbf{q})^2}, \qquad (6.52)$$

where the prime on the summation indicates that the $q = 0$ term is omitted. From eqn (6.35) this may be written in terms of the Lindhard response function as

$$U_{bs}^{(2)} = \tfrac{1}{2}\Omega \sum_q' |S(\mathbf{q})|^2 [\chi_0(\mathbf{q})/\varepsilon^2(\mathbf{q})] |v_{ps}^{ion}(\mathbf{q})|^2. \qquad (6.53)$$

However, summing over all the occupied energy levels has counted the coulomb interaction between any pair of electrons, $i$ and $j$, twice. This follows since

$$\int \rho_i(\mathbf{r}) V_j(\mathbf{r}) \, d\mathbf{r} + \int \rho_j(\mathbf{r}) V_i(\mathbf{r}) \, d\mathbf{r} = 2\left[ \int\int \rho_i(\mathbf{r}) [e^2/4\pi\varepsilon_0 |\mathbf{r} - \mathbf{r}'|] \rho_j(\mathbf{r}') \, d\mathbf{r} \, d\mathbf{r}' \right],$$

$$(6.54)$$

where $V_i$ and $V_j$ are the electrostatic contributions to the potential, which enter the Schrödinger equation from electrons, $i$ and $j$, respectively. We have written the coulomb interaction explicitly as $e^2/4\pi\varepsilon_0 |\mathbf{r} - \mathbf{r}'|$ on the right-hand side, in order to avoid the confusion that $e^2 = 2$ in atomic units. Thus, we must subtract off the total electrostatic energy of interaction between the electrons that has been *double-counted* in $U^{(2)}$. From the right-hand side of eqn (6.54), the second-order contribution may be written in $\mathbf{q}$-space as

$$\frac{1}{2\mathcal{N}} \sum_q' \delta\rho_q^*(8\pi/q^2)\delta\rho_q = \frac{1}{2\Omega} \sum_{q'} |S(\mathbf{q})|^2 \{[8\pi\chi_0^2(\mathbf{q})/q^2]/\varepsilon^2(q)\} |v_{ps}^{ion}(\mathbf{q})|^2, \quad (6.55)$$

since $\delta\rho_q = \chi(\mathbf{q}) V(\mathbf{q})$.

The so-called *band-structure* contribution to the total binding energy per atom is given by subtracting this double-counting term, eqn (6.55), from $U_{bs}^{(2)}$. Using eqn (6.26) we have

$$U_{bs} = \tfrac{1}{2}\Omega \sum_q' |S(\mathbf{q})|^2 [\chi(\mathbf{q})/\varepsilon(\mathbf{q})] |v_{ps}^{ion}(\mathbf{q})|^2, \qquad (6.56)$$

where we have replaced the Lindhard response function, $\chi_0$, by the exchange-correlation enhanced response function $\chi(\mathbf{q})$ (cf eqn (6.36)). By convention

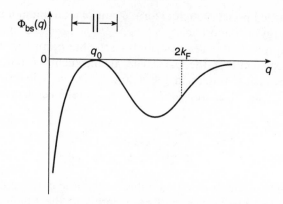

**Fig. 6.6** The wave-vector dependence of the energy-wavenumber characteristic, $\Phi_{bs}(q)$, which has a node at $q_0$ and a weak logarithmic singularity in its slope at $q = 2k_F$. Also shown are a set of degenerate cubic reciprocal lattice vectors that are centred on $q_0$. A tetragonal distortion would lift their degeneracy away from the node at $q_0$ as shown, thereby lowering the band-structure energy. (After Heine and Weaire (1970).)

this simplifies as

$$U_{bs} = \sum_{\mathbf{q}}' |S(\mathbf{q})|^2 \Phi_{bs}(\mathbf{q}) \tag{6.57}$$

by defining the *energy-wave number characteristic* $\Phi_{bs}(\mathbf{q})$ by

$$\Phi_{bs}(\mathbf{q}) = \tfrac{1}{2}\Omega[\chi(\mathbf{q})/\varepsilon(\mathbf{q})]|v_{ps}^{ion}(\mathbf{q})|^2 \tag{6.58}$$

The *energy-wave-number characteristic*, $\Phi_{bs}(\mathbf{q})$, depends only on the density of the free-electron gas and the nature of the pseudopotential core but not on the structural arrangement of the atoms. Its behaviour as a function of the wave vector, $\mathbf{q}$, is illustrated in Fig. 6.6, where we see that it vanishes at $q_0$ as expected. It also has a weak logarithmic singularity in its slope at $q = 2k_F$.

The total binding energy per atom of the NFE metal can then be written as the sum of the three terms:

$$U = ZU_{eg} + U_{es} + U_{bs}, \tag{6.59}$$

which extends eqn (5.59) to include the second-order *band-structure* contribution. The electrostatic contribution $U_{es}$ is formally given by

$$U_{es} = U_{Madelung} + U_{core}, \tag{6.60}$$

where

$$U_{Madelung} = -\alpha Z^2/R_{WS}, \tag{6.61}$$

and

$$U_{core} = \lim_{q \to 0} \{Z[v_{ps}^{ion}(\mathbf{q}) - 8\pi Z/\Omega q^2]\}. \tag{6.62}$$

**Table 6.1** The Madelung constant $\alpha$ for various structures

| Structure | $\alpha$ |
|---|---|
| bcc | 1.79186 |
| fcc | 1.79175 |
| hcp (ideal) | 1.79168 |
| simple hexagonal | 1.77464 |
| simple cubic | 1.76012 |

The Madelung energy is the electrostatic energy of point ions with charge, $Ze$, that are immersed in a neutralizing uniform distribution of electrons. We had evaluated this term within the Wigner–Seitz sphere approximation in eqn (5.62), obtaining the value of the Madelung constant, $\alpha = (3 - 1.2) = 1.8$. In practice, close-packed metals have Madelung constants that are about 0.4% smaller than this as can be seen from Table 6.1. The core energy may be evaluated explicitly for Ashcroft empty-core pseudopotentials, since substituting eqn (5.56) into eqn (6.61) we then have

$$U_{\text{core}} = \lim_{q \to 0} \left\{ \frac{8\pi Z^2}{\Omega} \left[ -\frac{(1 - \frac{1}{2}q^2 R_c^2 + \cdots)}{q^2} + \frac{1}{q^2} \right] \right\}, \tag{6.63}$$

so that

$$U_{\text{core}} = \frac{3Z^2}{R_{\text{ws}}} \left( \frac{R_c}{R_{\text{ws}}} \right)^2, \tag{6.64}$$

which is just the core contribution we had found earlier in eqn (5.62).

Second-order perturbation theory has allowed us to write the band-structure energy of a perfect crystal as a sum over reciprocal lattice vectors, since the structure factor vanishes unless $\mathbf{q} = \mathbf{G}$ (cf eqn (6.59)). Thus, within the *reciprocal lattice representation* we have

$$U_{\text{bs}} = \sum_{\mathbf{G}} |S(\mathbf{G})|^2 \Phi_{\text{bs}}(\mathbf{G}). \tag{6.65}$$

This demonstrates immediately the importance of the location of the first few reciprocal lattice vectors with respect to the node at $q_0$ in the energy-wavenumber characteristic, $\Phi_{\text{bs}}(\mathbf{q})$, which is illustrated in Fig. 6.6. Vectors lying at $q_0$ would contribute nothing to the band-structure energy, which is consistent with zero band gaps opening up at the appropriate zone boundary. We found earlier that the $\mathbf{G}(111)$ and $\mathbf{G}(200)$ reciprocal lattice vectors of aluminium lay just to the right of $q_0$ in Fig. 5.12, so that small negative contributions to the band-structure energy in eqn (6.65) are obtained.

However, on going down group III in the periodic table from Al → Ga → In we see from Table 6.2 that $\mathbf{G}(200)$ moves closer to $q_0$ and $\mathbf{G}(111)$, in fact,

Table 6.2 Values of $G/q_0$ for fcc aluminium, gallium, and indium. (From Heine and Weaire (1970).)

|  | Al | Ga | In |
|---|---|---|---|
| $G(111)/q_0$ | 1.04 | 0.94 | 0.93 |
| $G(200)/q_0$ | 1.20 | 1.09 | 1.08 |

passes right through $q_0$ to the other side. This is reflected by a change in the symmetry of the state at the bottom of the gap at $L = (2\pi/a)\,(1, 1, 1)$ on the Brillouin zone boundary. Whereas this state has the p-like symmetry, $L_{2'}$, in aluminium (cf Fig 5.9), it has the s-like symmetry, $L_1$, in fcc gallium and indium. This behaviour in $G/q_0$ causes the reciprocal lattice vectors across the first Brillouin zone in fcc gallium and indium to lie closer to $q_0$ on average than those for fcc aluminium. It is, therefore, not implausible that the gallium and indium lattices distort from fcc in order to gain a marked lowering in their band-structure energy as their distorted reciprocal lattice vectors move away on either side from the node at $q_0$. A cubic to tetragonal distortion is illustrated schematically in Fig. 6.6, as is indeed observed for indium with $c/a = 1.08$.

In practice, the degree of distortion is obtained by a balance between the electrostatic Madelung energy, which increases with distortion, and the band-structure energy, which decreases with distortion. The real-space representation presented in the next section allows these two opposing contributions to be combined together into a single term, thereby facilitating an understanding of the observed structural trends within the elemental sp-valent metals.

## 6.5. The real-space representation

The total binding energy can be rewritten as a sum over real space rather than $\mathbf{q}$-space vectors by changing the order of their summation in $U_{bs}$. From eqn (6.57) the band-structure energy is given by

$$U_{bs} = \sum_{\mathbf{q}} S^*(\mathbf{q})S(\mathbf{q})\Phi_{bs}(\mathbf{q}) - \lim_{\mathbf{q}\to 0}\Phi_{bs}(\mathbf{q}), \qquad (6.66)$$

where the $\mathbf{q} = 0$ term has been included explicitly in the summation. Substituting in for the structure factor from eqn (6.47) and separating out the $\mathbf{R}_i = \mathbf{R}_j$ terms we have

$$U_{bs} = \frac{1}{2\mathcal{N}}\sum_{i\neq j}\left\{\frac{2}{\mathcal{N}}\sum_{\mathbf{q}}\Phi_{bs}(\mathbf{q})\,e^{i\mathbf{q}\cdot(\mathbf{R}_i - \mathbf{R}_j)}\right\} + \frac{1}{\mathcal{N}}\sum_{\mathbf{q}}\Phi_{bs}(\mathbf{q}) - \lim_{\mathbf{q}\to 0}\Phi_{bs}(\mathbf{q}). \quad (6.67)$$

This allows us to define the band-structure contribution to a pair potential, namely

$$\Phi_{bs}(R) = \frac{2}{\mathcal{N}} \sum_q \Phi_{bs}(q)\, e^{iq \cdot R}. \tag{6.68}$$

Thus, we can write

$$U_{bs} = \frac{1}{2\mathcal{N}} \sum_{i \neq j} \Phi_{bs}(R_{ij}) + \tfrac{1}{2}\Phi_{bs}(R = 0) - \lim_{q \to 0} \Phi_{bs}(q). \tag{6.69}$$

The Madelung energy, eqn (6.61), can also be expressed as a pairwise sum over coulomb interactions between the point ions, plus a $q \to 0$ contribution arising from the electron–ion and electron–electron interactions. Grouping this together with the band-structure contribution we have

$$U_{bs} + U_{Madelung} = \frac{1}{2\mathcal{N}} \sum_{i \neq j} \Phi(R_{ij}) + \tfrac{1}{2}\Phi_{bs}(R = 0) - \lim_{q \to 0}\left[ \Phi_{bs}(q) + \frac{1}{2}\frac{8\pi Z^2}{q^2\Omega} \right], \tag{6.70}$$

where the interatomic pair potential $\Phi(R)$ is given by the sum of the ion–ion and band-structure terms, namely

$$\Phi(R) = 2Z^2/R + \Phi_{bs}(R). \tag{6.71}$$

In 1974 Finnis showed that the $q \to 0$ limit could be evaluated directly by using the compressibility sum rule of the free-electron gas, which relates the long wavelength behaviour of the dielectric constant to its compressibility. He found that

$$\lim_{q \to 0}\left[ \Phi_{bs}(q) + \frac{1}{2}\frac{8\pi Z^2}{q^2\Omega} \right] = \tfrac{1}{2}Z\Omega B_{eg} + U_{core}, \tag{6.72}$$

where $B_{eg} = \Omega d^2 U_{eg}/d\Omega^2$ is the electron–gas bulk modulus or inverse compressibility, and $U_{core}$ is the core contribution to the binding energy, eqn (6.61).

Thus, *the total binding energy per atom of a NFE metal can be expressed in a physically transparent form, as the sum of a volume-dependent contribution and a pair-potential contribution* in a manner that is reminiscent of the semi-empirical embedded atom potential of eqn (5.68). It follows from eqs (6.59)–(6.72) that

$$U = U_0(\Omega) + \frac{1}{2\mathcal{N}} \sum_{i \neq j} \Phi(R_{ij}, \Omega), \tag{6.73}$$

where

$$U_0(\Omega) = Z(U_{eg} - \tfrac{1}{2}\Omega B_{eg}) + \tfrac{1}{2}\Phi_{bs}(R = 0). \tag{6.74}$$

The electron–gas term in eqn (6.74) is given from eqn (5.60) by

$$U_{eg} - \tfrac{1}{2}\Omega B_{eg} = 0.982/r_s^2 - 0.712/r_s - (0.110 - 0.031 \ln r_s). \quad (6.75)$$

The band-structure term is given from eqs (6.68) and (6.58) by

$$\tfrac{1}{2}\Phi_{bs}(\mathbf{R} = 0) = \frac{1}{2}\left[\frac{\Omega}{\mathscr{N}} \sum_{\mathbf{q}} \delta\rho(\mathbf{q}) v_{ps}^{ion}(\mathbf{q})\right] \quad (6.76)$$

as $\delta\rho(\mathbf{q}) = \chi(\mathbf{q}) v_{ps}(\mathbf{q})$. This contribution is, therefore, one half of the electro-static interaction between an ion and its own screening cloud, so that it represents the binding energy of a screened pseudoatom. This expression is consistent with the virial theorem, where the equilibrium binding energy of all the electrons in a free atom is just one half of the total potential energy. It may be approximated by the simple expression (Finnis (1974))

$$\tfrac{1}{2}\Phi_{bs}(\mathbf{R} = 0) \simeq - Z^2 q_0/\pi = -\tfrac{1}{2}Z^2/R_c \quad (6.77)$$

where the last equality holds for the Ashcroft empty-core pseudopotential from eqn (5.57). We see from Table 6.3 that eqn (6.77) reflects the large variation in the binding energy of the pseudoatoms in going across the period from Na $\rightarrow$ Mg $\rightarrow$ Al.

The interatomic pair potential $\Phi(R \neq 0)$ in eqn (6.73) represents the electrostatic interaction between an ion and a second ion and its screening cloud some distance, $R$, away. From eqn (6.71) it is given by

$$\Phi(\mathbf{R}) = \frac{2Z^2}{R} + \frac{2\Omega}{(2\pi)^3} \int \Phi_{bs}(\mathbf{q}) \, e^{i\mathbf{q}\cdot\mathbf{R}} \, d\mathbf{q}, \quad (6.78)$$

**Table 6.3** Contributions to the binding energy (in Ry per atom) of sodium, magnesium, and aluminium within the second order real-space representation, eqn (6.73), using Ashcroft empty-core pseudopotentials. $U_{eg}^{(2)}$ is defined by eqn (6.75). The numbers in brackets correspond to the simple expression, eqn (6.77), for $\tfrac{1}{2}\Phi_{bs}(R = 0)$ and to the experimental values of the binding energy and negative cohesive energy respectively. (From Hafner (1987).)

| | $R_c$(au) | $ZU_{eg}^{(2)}$ | $\tfrac{1}{2}\Phi_{bs}(0)$ | $U_0(\Omega)$ | $\frac{1}{2\mathscr{N}}\sum\Phi(R)$ | $U$ | $-U_{coh}$ |
|---|---|---|---|---|---|---|---|
| Na | 1.71 | $-0.19$ | $-0.29$ | $-0.48$ | $-0.02$ | $-0.49$ | — |
| | | | $(-0.29)$ | | | $(-0.46)$ | $(-0.08)$ |
| Mg | 1.31 | $-0.42$ | $-1.49$ | $-1.91$ | $-0.02$ | $-1.92$ | — |
| | | | $(-1.52)$ | | | $(-1.78)$ | $(-0.11)$ |
| Al | 1.11 | $-0.61$ | $-3.91$ | $-4.52$ | $+0.03$ | $-4.49$ | — |
| | | | $(-4.05)$ | | | $(-4.16)$ | $(-0.25)$ |

where we have replaced the summation in eqn (6.68) by an integral in the usual way. Because the energy-wave-number characteristic, $\Phi_{bs}$, depends only on the magnitude of $\mathbf{q}$ (cf eqn (6.58)), the integration over the solid angle may be carried out in eqn (6.78) to yield

$$\Phi(R) = \frac{2Z^2}{R} + \frac{\Omega}{\pi^2} \int_0^\infty \Phi_{bs}(q)\left(\frac{\sin qR}{qR}\right)q^2\,dq. \tag{6.79}$$

Defining the normalized ion-core pseudopotential matrix element by

$$\hat{v}_{ps}^{ion}(q) = [(\Omega q^2)/8\pi Z]v_{ps}^{ion}(q), \tag{6.80}$$

and using eqn (6.58) for $\Phi_{bs}(q)$, the interatomic pair potential may be written

$$\Phi(R) = \frac{2Z^2}{R}\left\{1 - \frac{2}{\pi}\int_0^\infty \left[\frac{\varepsilon(q)-1}{\varepsilon(q)}\right][v_{ps}^{ion}(q)]^2\,\frac{\sin qR}{q}\,dq\right\}, \tag{6.81}$$

which can be regrouped as

$$\Phi(R) = \frac{2Z^2}{R}\left\{1 - \frac{2}{\pi}\int_0^\infty [\hat{v}_{ps}^{ion}(q)]^2\,\frac{\sin qR}{q}\,dq\right\}$$
$$+ \frac{2Z^2}{R}\left\{\frac{2}{\pi}\int_0^\infty [\varepsilon(q)]^{-1}[\hat{v}_{ps}^{ion}(q)]^2\,\frac{\sin qR}{q}\,dq\right\}. \tag{6.82}$$

The term within the first set of curly brackets vanishes as can be seen by converting the integral to the usual semi-circular path within the complex plane and noting that $\hat{v}_{ps}^{ion}(q=0) = 1$ (cf eqn (5.56)).

Thus, the pair potential can be expressed by the *single* term

$$\Phi(R) = \frac{4Z^2}{\pi R}\int_0^\infty \hat{v}_{ps}(q)\hat{v}_{ps}^{ion}(q)\,\frac{\sin qR}{q}\,dq, \tag{6.83}$$

where $\hat{v}_{ps}(q) = \hat{v}_{ps}^{ion}(q)/\varepsilon(q)$. As expected, this integral represents the interaction between a bare ion and a screened ion or pseudoatom some distance, $R$, away. The very weak logarithmic singularity in the slope of the response function at $q = 2k_F$ leads to the *asymptotic* Friedel oscillations

$$\Phi(R) \sim A\cos 2k_F R/R^3, \tag{6.84}$$

which are similar to those already observed in the screening cloud of eqn (6.41). Figure 6.7 shows the interatomic pair potentials for sodium, magnesium, and aluminium at their respective equilibrium volumes. They were computed using *non-local* pseudopotentials whose use requires a generalization of the formalism presented above. The oscillatory nature of the pair potential with respect to the first few nearest-neighbour shells of

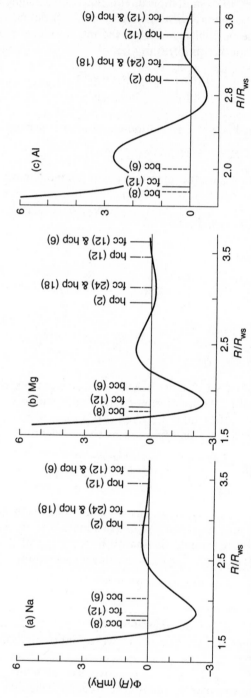

**Fig. 6.7** Interatomic pair potentials for (a) Na, (b) Mg, and (c) Al as a function of the interatomic separation in units of the Wigner–Seitz radius $R_{WS}$. The positions of the fcc first and bcc first and second nearest neighbours are marked. Since fcc and ideal hcp structures have identical first and second nearest neighbours, their relative structural stability is determined by the more distant neighbours marked in the figures. (After McMahan and Moriarty (1983).)

atoms is clearly observed. We see that, whereas sodium and magnesium have an attractive pair potential of about $-2$mRy at the nearest-neighbour distance, aluminium has a repulsive pair potential of about $+2$mRy. Thus, just as was predicted by the *local* Ashcroft empty-core pseudopotentials in Table 6.3, the pair potential contribution to the total binding energy in aluminium is repulsive not attractive as for sodium and magnesium.

Table 6.3 illustrates the very important fact that the structurally dependent part of the energy contributes only about 1% to the total binding energy of sp-valent metals. This is not unexpected given the similarity between their densities of states and that of a free-electron gas. The largest contribution is the energy of the screened ion or pseudoatom, which varies approximately as $-Z^2/2R_c$. We should note that the cohesive energy cannot be predicted by second-order perturbation theory since this would require a prediction of the binding energy of the free atom. This, of course, cannot be approximated as a weak perturbation of a jellium sphere. The eigenspectrum of a free atom and a jellium sphere are totally different as is reflected by their very different magic numbers or closed-shell occupancies (cf §§2.6 and 5.2).

However, second-order perturbation theory can compare one bulk situation with another, since errors in the absolute prediction of the binding energy (usually residing in the energy of the individual pseudoatoms, $\frac{1}{2}\Phi_{bs}(\mathbf{R} = 0)$) often cancel out. Thus, for example, the heat of formation of a binary AB compound can be evaluated reliably since $\Delta H$ compares the binding energy of the AB compound with that of the A and B elemental metals, that is

$$\Delta H = U_{AB} - \tfrac{1}{2}(U_A + U_B). \tag{6.85}$$

This can be written within the real-space representation as

$$\Delta H = \Delta H_{eg} + \Delta H_{pa} + \Delta H_{struct}, \tag{6.86}$$

where the electron gas, pseudoatom, and structure-dependent contributions correspond to the three different terms resulting from eqs (6.73) and (6.74). Assuming that the binding energy of the pseudoatoms is given by eqn (6.77) and that the structure-dependent contribution is small and can be neglected, the heat of formation will be determined by the electron-gas term alone. Taking the equilibrium atomic volume, $\Omega$, of the compound to be equal to the average elemental atomic volumes $\frac{1}{2}(\Omega_A + \Omega_B)$ by Zen's law, we find from eqn (6.75) that the electron–gas contribution can be written to second order as

$$\Delta H_{eg} = Z(-1.228 + 0.356r_s + 0.031r_s^2)[\Delta(1/r_s)]^2 \tag{6.87}$$

where $Z$ is the average valence, $\frac{1}{2}(Z_A + Z_B)$.

We may express this in terms of the more commonly used variable, the electron density, as from eqn (2.41) $\rho^{1/3} = (3/4\pi)^{1/3}(1/r_s)$. The heat of formation in eV per atom is then given by

$$\Delta H \approx \Delta H_{eg} = Z(-43.39 + 7.81\rho^{-1/3} + 0.17\rho^{-2/3})(\Delta\rho^{1/3})^2. \tag{6.88}$$

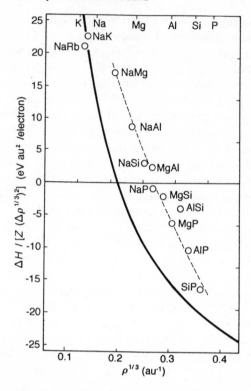

**Fig. 6.8** The normalized heat of formation, $\Delta H/[Z(\Delta\rho^{1/3})^2]$, as a function of the average cube root of the electron density, $\rho^{1/3}$, for sp-valent AB compounds. The solid curve is the electron–gas contribution, eqn (6.88). The open circles are the self-consistent local density approximation predictions for the CsCl lattice. (From Pettifor and Gelatt (1983).)

The three terms inside the brackets are the kinetic, exchange and correlation contributions respectively. Mixing together two electron gases of equilibrium densities $\rho_A$ and $\rho_B$ respectively to form a new electron gas of average density, $\rho$, lowers the kinetic energy but raises the exchange and correlation energies. Figure 6.8 shows that this electron–gas contribution indeed reproduces the broad behaviour in the heats of formation of simple metal compounds. In particular, we note that $\Delta H$ changes sign from positive at low densities (where the exchange and correlation dominates) to negative at high densities (where the kinetic energy dominates).

## 6.6. Structural trends

The beauty of the real-space representation is that it separates out the very small structure-dependent contribution to the total binding energy as a single sum over pair potentials. As illustrated in Fig. 6.7, these metallic pair potentials oscillate with respect to the nearest-neighbour positions, so that an understanding of relative structural stability requires an understanding of the origin of these oscillations. These near-neighbour oscillations are *not* due to the weak logarithmic singularity *at* $2k_F$, since this determines the very long range asymptotic behaviour of eqn (6.84). Instead they are determined by the *overall* shape of the Lindhard response function.

This can be demonstrated by replacing the Lindhard function by a rational polynomial, which has the correct low and high $q$ behaviour, and is such that it passes through $\frac{1}{2}\chi_{TF}$ at $q = 2k_F$. Such a rational polynomial, correct to third order in both $q^2$ and $1/q^2$ is given by the dashed curve in Fig. 6.3. It has the form (Pettifor and Ward (1984))

$$\frac{\chi_0(\eta)}{\chi_{TF}} =$$

$$\frac{1 + b\eta^2 + \left(\dfrac{61}{315} + \dfrac{4b}{15} - \dfrac{3c}{35}\right)\eta^4 + \left(\dfrac{16}{63} + \dfrac{b}{3} - \dfrac{8c}{35}\right)\eta^6 + c\eta^8}{1 + \left(b + \dfrac{1}{3}\right)\eta^2 + \left(\dfrac{13}{35} + \dfrac{3b}{5} - \dfrac{3c}{35}\right)\eta^4 + \left(\dfrac{3}{7} + \dfrac{3b}{5} - \dfrac{9c}{35}\right)\eta^6 + \left(\dfrac{16}{21} + b - \dfrac{9c}{7}\right)\eta^8 + 3c\eta^{10}},$$

(6.89)

where the arbitrary constants $b$ and $c$ were chosen to minimize the rms error, namely $b = -0.5395$ and $c = 0.3333$. This rational polynomial was obtained by expanding the logarithms in the Lindhard function, eqn (6.39), to yield

$$\chi_0(\eta)/\chi_{TF} = \begin{cases} 1 - \displaystyle\sum_{n=1}^{\infty} \dfrac{\eta^{2n}}{(2n-1)(2n+1)} & \text{for } \eta < 1 \\[4mm] \displaystyle\sum_{n=1}^{\infty} \dfrac{(1/\eta)^{2n}}{(2n-1)(2n+1)} & \text{for } \eta > 1. \end{cases}$$

(6.90)

It is easy to verify that a binomial expansion of eqn (6.89) about $\eta = 0$ or about $1/\eta = 0$ yields eqn (6.90) correct to the third order in $\eta^2$ or $1/\eta^2$ respectively. Clearly, the fit of the Lindhard function can be made as accurate as one pleases by including further terms in the numerator and denominator of eqn (6.89).

The integral defining the pair potential in eqn (6.83) may now be evaluated directly. The poles of the inverse dielectric function, $\varepsilon^{-1}(q)$, are first found by substituting eqn (6.89) into eqn (6.36) and writing

$$\varepsilon^{-1}(q) = \sum_{n=1}^{6} D_n q^2/(q^2 - q_n^2).$$

(6.91)

The weights, $D_n$, and poles, $q_n$, are, in general, complex, so that we may write

$$D_n = d_n \, e^{i\delta_n}$$

(6.92)

and

$$q_n = k_n + i\kappa_n. \tag{6.93}$$

Substituting eqn (6.91) into eqn (6.83) and performing conventional contour integration we find

$$\Phi(R) = (2Z^2/R) \sum_{n=1}^{6} D_n[\hat{v}_{ps}^{ion}(q_n)]^2 \, e^{iq_n R}. \tag{6.94}$$

At normal metallic densities, the six poles are complex, so that we may assign $q_4^2 = (q_1^2)^*$, $q_5^2 = (q_2^2)^*$, and $q_6^2 = (q_3^2)^*$.

*The pair potential may thus be approximated by the sum of three damped oscillatory terms*, namely

$$\Phi(R) = (2Z^2/R) \sum_{n=1}^{3} A_n \cos(k_n R + \alpha_n) \, e^{-\kappa_n R}, \tag{6.95}$$

where the amplitude, $A_n$, is given by

$$A_n = 2d_n|\hat{v}_{ps}^{ion}(q_n)|^2, \tag{6.96}$$

and the phase, $\alpha_n$, is given by

$$\alpha_n = \delta_n + 2 \arg \hat{v}_{ps}^{ion}(q_n). \tag{6.97}$$

The wave vector, $k_n$, and the screening length, $1/\kappa_n$, depend only on the density of the free-electron gas through the poles of the approximated inverse dielectric response function, whereas the amplitude, $A_n$, and the phase shift, $\alpha_n$, depend also on the nature of the ion-core pseudopotential through eqs (6.96) and (6.97). For the particular case of the Ashcroft empty-core pseudopotential, where $\hat{v}_{ps}^{ion}(q) = \cos qR_c$, the modulus and phase are given explicitly by

$$|\hat{v}_{ps}^{ion}(q_n)| = \frac{1}{\sqrt{2}} (\cos 2k_n R_c + \cosh 2\kappa_n R_c)^{1/2} \tag{6.98}$$

and

$$\tan[\arg \hat{v}_{ps}^{ion}(q_n)] = -\tan k_n R_c \tanh \kappa_n R_c. \tag{6.99}$$

The resultant pair potentials for sodium, magnesium, and aluminium are illustrated in Fig. 6.9 using Ashcroft empty-core pseudopotentials. We see that all three metals are characterized by a repulsive hard-core contribution, $\Phi_1(R)$ (short-dashed curve), an attractive nearest-neighbour contribution, $\Phi_2(R)$ (long-dashed curve), and an oscillatory long-range contribution, $\Phi_3(R)$ (dotted curve). The appropriate values of the inter-atomic potential parameters $A_n$, $\alpha_n$, $k_n$, and $\kappa_n$ are listed in Table 6.4. We observe that the total pair potentials reflect the characteristic behaviour of the more accurate *ab initio* pair potentials in Fig. 6.7 that were evaluated using non-local pseudopotentials. We should note, however, that the values taken for the Ashcroft empty-core radii for Na, Mg, and Al, namely $R_c = 1.66, 1.39$, and

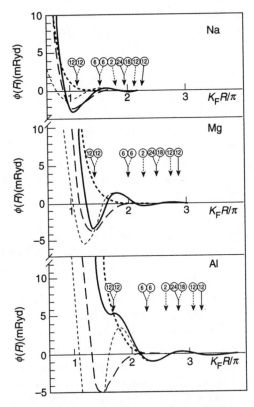

**Fig. 6.9** The analytic pair potentials (———) for sodium, magnesium, and aluminium. The short-range (– – –), medium-range (— —), and long-range (- - -), oscillatory terms are also given. The full (dotted) arrows mark the positions of the first four (five) nearest-neighbour shells in fcc (ideal hcp). (From Pettifor and Ward (1984).)

1.12 au, are different from those obtained by fitting to the equilibrium atomic volumes in Table 5.2, namely $R_c = 1.75$, 1.31, and 1.11 au respectively. It is not possible with this one-parameter Ashcroft potential to fit both the observed equilibrium volume and crystal structure. This could be achieved by using a smooth two-parameter parabolic model potential rather than the discontinuous Ashcroft potential, but the analytic expressions are too complicated for the simple treatment attempted in this book.

The three contributions to the pair potential can be analysed in more detail.

(i) The *short-range* potential $\Phi_1(R)$ can be approximated by a Yukawa screened potential, namely

$$\Phi_1(R) \approx (2Z^2/R)A \, e^{-\kappa_1(R-2R_c)}. \tag{6.100}$$

**Table 6.4** Pair potential parameters for sodium, magnesium, and aluminium at their equilibrium volumes. (From Pettifor and Ward (1984).)

|  | Na | | | Mg | | | Al | | |
|---|---|---|---|---|---|---|---|---|---|
| $n$ | 1 | 2 | 3 | 1 | 2 | 3 | 1 | 2 | 3 |
| $k_n/2k_F$ | 0.291 | 0.715 | 0.958 | 0.224 | 0.664 | 0.958 | 0.156 | 0.644 | 0.958 |
| $\kappa_n/2k_F$ | 0.897 | 0.641 | 0.271 | 0.834 | 0.675 | 0.277 | 0.793 | 0.698 | 0.279 |
| $A_n$ | 1.961 | 0.806 | 0.023 | 5.204 | 1.313 | 0.033 | 7.964 | 1.275 | 0.030 |
| $\alpha_n/\pi$ | −0.294 | −0.750 | −0.995 | −0.401 | 0.932 | 0.499 | −0.441 | 0.832 | 0.431 |

This follows directly from eqs (6.95)–(6.99) using $\cosh x \approx \frac{1}{2} \exp x$ for large values of $x$. The prefactor $A$ has absorbed the oscillatory contribution $\cos(k_1 R + \alpha_1)$ in eqn (6.95) since it varies only slowly with distance due to the small value of $k_1$ (cf Table 6.4). At high metallic densities $\kappa_1 \rightarrow \kappa_{TF}$. Thus, the short-range potential reflects the *repulsive* interaction between two ion cores of radius $R_c$, the range of the repulsion being determined by the Thomas–Fermi screening length $\lambda_{TF} = \kappa_{TF}^{-1}$.

(ii) The *medium-range* potential $\Phi_2(R)$ is characterized by an *attractive* minimum close to the nearest neighbour separation for sodium, magnesium and aluminium in Fig. 6.9. Differentiating $\Phi_2(R)$ with respect to distance neglecting its weak inverse $R$ dependence, the position of the minimum $R_{\min}$ satisfies

$$R_{\min} = -[\tan^{-1}(\kappa_2/k_2) + \alpha_2]/k_2. \qquad (6.101)$$

The core dependence of $R_{\min}$ can then be found by differentiating eqn (6.101) with respect to $R_c$, namely

$$\frac{\partial R_{\min}}{\partial R_c} = -\frac{1}{k_2}\frac{\partial \alpha_2}{\partial R_c} = -\frac{2}{k_2}\frac{\partial}{\partial R_c}[\arg \hat{v}_{ps}^{ion}(q_2)] \qquad (6.102)$$

since $k_2$, $\kappa_2$, and $\delta_2$ depend only on the electron density and not on the core radius $R_c$. But it follows from eqn (6.99) that

$$\arg \hat{v}_{ps}^{ion}(q_2) \approx -k_2 R_c \qquad (6.103)$$

since $\tanh \kappa_2 R_c \approx 1$ within 10% at equilibrium metallic densities. Thus, substituting eqn (6.103) into (6.102) we have

$$\frac{\partial R_{\min}}{\partial R_c} \approx 2. \qquad (6.104)$$

This is consistent with Hafner and Heine (1983) who found *numerically* that the minimum in the total potential satisfied

$$R_{min} = 2.0R_c + 3.3 \pm 0.1 \qquad (6.105)$$

in atomic units. This minimum only arises because of the oscillatory wave-like character of the screening clouds in Fig. 6.4. It is not predicted within the Thomas–Fermi approximation.

(iii) The *long-range* potential $\Phi_3(R)$ dominates soon after the first minimum in the total potential. It is characterized by a wavevector $k_3$ that is to within 4% of the wavevector $2k_F$ that describes the very long-range asymptotic Friedel behaviour in eqn (6.84). Using the high-density limiting values of $k_3$ and $\kappa_3$ (cf Table 6.4), the long-range potential takes the form

$$\Phi_3(R) = (2Z^2/R)A_3 \cos[0.958(2k_F R) + \alpha_3] e^{-0.574k_F R}. \qquad (6.106)$$

The argument $k_F R$ is independent of electron density or atomic volume since it can be written from eqn (2.40) in terms of the Wigner–Seitz radius $R_{WS}$ as

$$k_F R = (9\pi/4)^{1/3} R/r_s = (9\pi/4)^{1/3} Z^{1/3} R/R_{WS}. \qquad (6.107)$$

Thus, whereas the asymptotic Friedel oscillations in eqn (6.84) have their *phase* fixed with respect to the underlying lattice for a given valence $Z$, the oscillations of the long-range potential are electron density or atomic volume sensitive through the phase shift $\alpha_3$ which from eqs (6.97) and (6.99) is given by

$$\alpha_3 = \delta_3 - 2 \tan^{-1}[\tan(3.68R_c/r_s) \tanh(1.10R_c/r_s)]. \qquad (6.108)$$

Hence, a reduction in volume will cause the phase shift to decrease as shown in Fig. 6.10. This decrease in $\alpha_3$ causes the long-range oscillations to *move out* with respect to the nearest neighbour positions in Fig. 6.7 and Fig. 6.9, thereby causing a possible change in the relative stability of different competing metallic structure types.

This is illustrated by considering the relative stability of the fcc, bcc, and hcp lattices of sodium, magnesium, and aluminium with respect to a decrease in atomic volume about their equilibrium volumes $\Omega_0$. The upper panel of Fig. 6.11 shows the energy difference between bcc and fcc (full curve) and hcp and fcc (dashed curve) as a function of the relative atomic volume $\Omega/\Omega_0$ that are obtained by summing over the *ab initio* interatomic potentials displayed in Fig. 6.7. We see that they predict that at equilibrium sodium and magnesium will be hcp but aluminium will be fcc. (At very low temperatures sodium, in fact, takes the samarium structure type with its mixed chh stacking sequence (cf §1.2) and transforms to bcc at 5 K. This is not inconsistent with the very small energy differences found theoretically in the upper left-hand panel.) Since fcc and the ideal close-packed hcp lattices have the same twelve first and six second nearest neighbour distances their relative structural stability is determined by the third, fourth and further neighbours that are marked in Fig. 6.11. We can see immediately

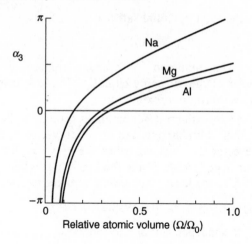

**Fig. 6.10** The phase shift, $\alpha_3$, of the long-range contribution to the pair potential for sodium, magnesium, and aluminium as a function of their relative atomic volume, $\Omega/\Omega_0$. (After Ward (1985).)

why magnesium is hcp but aluminium is fcc, since fcc magnesium and hcp aluminium would both have their twelve fourth nearest neighbours falling near maxima in the interatomic potentials. Under pressure all three metals are predicted to transform to bcc which has been confirmed experimentally for magnesium.

The volume dependence of the fcc, bcc, and hcp curves in the upper panel of Fig. 6.11 can be understood by using the *analytic* interatomic potentials of eqn (6.95). The middle panel demonstrates that they are capable of reproducing qualitatively the general features of the *ab initio* predictions. Remarkably, as is demonstrated by the lower panel, the behaviour of these structural energy difference curves is driven almost entirely by the long-range contribution to the pair potential $\Phi_3(R)$ since this oscillates with the shortest wavelength of the three contributions to $\Phi(R)$ (cf Table 6.4). We have already seen in Fig. 6.10 that the phase shift $\alpha_3$ decreases under pressure so that the long-range oscillations move out with respect to the nearest neighbour positions in Fig. 6.7 and Fig. 6.9. This can cause a change in structural stability that is most easily demonstrated by considering the close packed structures fcc and hcp. Focusing on the position of the twelve fifth nearest-neighbour hcp atoms in either figure, we see that as the oscillations move out under pressure the hcp phase in magnesium will initially have its stability increased with respect to fcc, whereas in sodium and magnesium it will be decreased. However, under still further compression, the phase shift $\alpha_3$ will eventually change by $\pi$ so that the relative hcp–fcc stability will reverse as

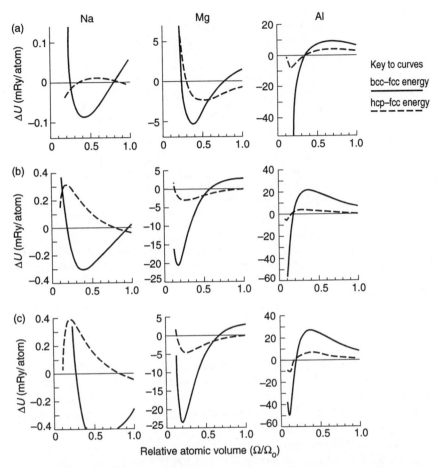

**Fig. 6.11** The structural-energy differences between bcc and fcc (full curves) and hcp and fcc (dashed curves) as a function of the relative atomic volume, $\Omega/\Omega_0$, for sodium, magnesium, and aluminium. The curves in the upper panel (a) were predicted by Moriarty and McMahan (1982) using their first principles interatomic potentials. The curves in the middle and lower panels (b) and (c) were predicted by Pettifor and Ward (1984) using three terms ($\Phi_1 + \Phi_2 + \Phi_3$) and one term $\Phi_3$ respectively in their analytic interatomic potentials.

found in Fig. 6.11. The behaviour of the bcc–fcc curves, on the other hand, is driven by the competition between the fourteen first and second nearest-neighbour contributions in bcc and the twelve first nearest-neighbour contributions in fcc (cf. question 6.1 in the Problems section at the end of the book).

Figure 6.12 shows the structure map, $(Z, \alpha_3)$, that is predicted using the

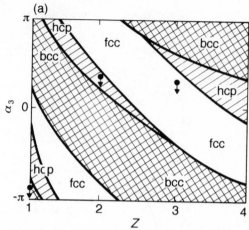

**Fig. 6.12** The structure map, $(Z, \alpha_3)$, that is predicted using the long-range pair potential, $\Phi_3(R)$. The three dots indicate the values of the phase shifts for sodium, magnesium, and aluminium corresponding to $Z = 1$, 2, and 3 respectively, the arrows indicating the direction the phase shift changes under pressure. (After Wyatt (1991).)

long-range pair potential, $\Phi_3(R)$, from eqn (6.100). In agreement with the lower panel of Fig. 6.11, we see that under compression sodium moves from bcc → fcc → hcp, magnesium moves from hcp → bcc → fcc, and aluminium moves from fcc → bcc → hcp. This map allows us to rationalize the structural trends that are observed down group III from Al → Ga → In → Tl, since these are conventional NFE metals with almost free-electron-like densities of states (cf Fig. 5.13). Experimentally it is found that $(2\pi/a)/q_0$ decreases down the group as is shown in Table 6.2. Therefore, from eqn (5.57) we expect $R_c/r_s$ to decrease and hence $\alpha_3$ to increase, so that the oscillations in $\Phi_3$ will *move in* with respect to the nearest-neighbour positions. This drives the stability from the fcc domain (of aluminium) to the nearby hcp domain (of thallium). However, as emphasized by Hafner and Heine (1983), this transition from fcc to hcp is accompanied by an instability of the close-packed lattices. This is due to one or more of their elastic shear constants becoming negative as the nearby *maximum* in $\Phi_3$ moves into the first nearest-neighbour position (cf the lower panel of Fig. 6.9). Thus, gallium and indium have *distorted* structures types with the twelve close-packed nearest neighbours splitting about the maximum in $\Phi_3(R)$.

The structural trends within the group II elements can only be understood by including the influence of the valence-d electrons explicitly through the use of non-local pseudopotentials. This is not unexpected considering our earlier discussion in §5.5 of their densities of states. Figure 6.13 shows the

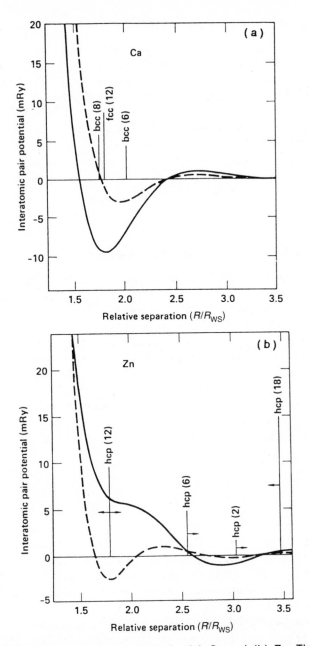

**Fig. 6.13** The interatomic pair potentials for (a) Ca and (b) Zn. The full and dashed curves correspond to including and excluding an explicit $l = 2$, 3d contribution, respectively. The positions of the fcc first and bcc first and second nearest neighbours are marked in the upper panel. The positions of ideal hcp first four nearest neighbours are given in the lower panel, the horizontal arrows indicating the directions these neighbours move as the $c/a$ axial ratio is increased above ideal. $R_{WS}$ is the Wigner–Seitz radius. (From Moriarty (1983).).

resultant interatomic pair potentials for calcium and zinc. We see that the influence of the *unfilled* 3d band in calcium causes the potential to deepen, thereby stabilizing the fcc structure, whereas the *filled* 3d band in zinc suppresses the nearest-neighbour minimum in $\Phi(R)$, thereby causing the ideal hcp lattice to relax to the much larger axial ratio of 1.86. This suppression of the local minimum becomes even more marked as one proceeds down group IIB, so that Hg is unstable with respect to all three lattices bcc, fcc, and ideal hcp, as illustrated in Fig. 6.14 for the case of $\Omega/\Omega_0 = 0.824$. The body-centred-tetragonal phase, $\beta$-Hg, is predicted to be most stable in agreement with experiment (cf Table 1.1).

## 6.7. Hume–Rothery electron phases

The most famous example of the crystal structure correlating with the average number of valence electrons per atom or band filling, $N$, is the Hume–Rothery alloy system of noble metals with sp-valent elements, such as Zn, Al, Si, Ge, and Sn. Assuming that Cu and Ag have a valence of 1, then the fcc $\alpha$-phase is found to extend to a value of $N$ around 1.38, the bcc $\beta$-phase to be stabilized around 1.48, the $\gamma$-phase around 1.62, and the hcp $\varepsilon$-phase around 1.75, as illustrated for the specific case of Cu–Zn alloys in Fig. 6.15. In 1936 Mott and Jones pointed out that the fcc and bcc electron per atom ratios correlate with the number of electrons required for a free-electron Fermi *sphere* first to make contact with the fcc and bcc Brillouin zone faces. The corresponding values of the Fermi vector, $k_F$, are given by

$$k_F = \begin{cases} \frac{1}{2}|\mathbf{G}(111)| = \sqrt{3}\pi/a & \text{for fcc} \\ \frac{1}{2}|\mathbf{G}(110)| = \sqrt{2}\pi/a & \text{for bcc} \end{cases} \tag{6.109}$$

Hence, since $k_F = (3\pi^2 N/\Omega)^{1/3}$ from eqn (2.40), the critical number of valence electrons, $N$, will be 1.36 and 1.48 for the fcc and bcc lattices respectively. These values, corresponding to $2k_F = |\mathbf{G}|$, lead to the asymptotic Friedel oscillations being in phase with the lattice, thereby giving rise to an additional stabilizing energy.

In 1937 Jones extended the model by including a realistic value for the copper energy gap at $L$, namely 4 eV, which had just been deduced from photoemission experiments. This is more than five times larger than the small energy gaps of NFE metals such as aluminium (cf Fig. 5.9). The large gap in copper arises from the orthogonality constraints imposed by the underlying valence 3d band. Jones found that this large gap caused the spherical free-electron Fermi surface to be so distorted that within his model it first makes contact with the fcc Brillouin zone boundary at $N = 1.04$ electrons.

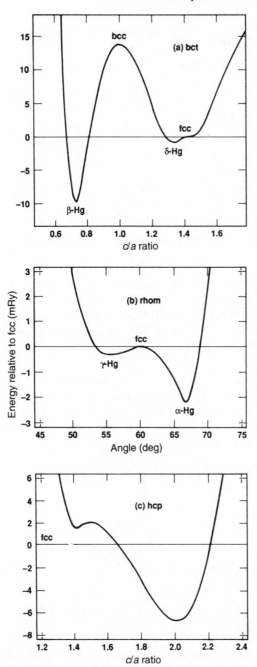

**Fig. 6.14** Relative binding energies of Hg at $\Omega/\Omega_0 = 0.824$ for three major structural families: (a) body-centred tetragonal, (b) simple rhombohedral, and (c) hexagonal close packed. (From Moriarty (1988).)

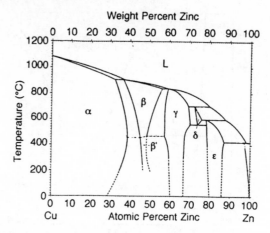

**Fig. 6.15** The Cu–Zn phase diagram. (After Massalski (1986).)

(Experimentally, it has *already* made contact in copper corresponding to $N = 1.00$ electrons). The resulting fcc and bcc densities of states look very similar to those for beryllium and lithium in Fig. 5.13 because Jones neglected the explicit presence of the copper 3d band. Comparing the fcc and bcc band energies, Jones found that the fcc lattice was indeed the more stable for $1 \leq N < 1.43$.

Figure 6.16 shows recent results of a Jones-type analysis of the stability of Cu–Zn alloys within the rigid-band approximation. This latter approximation assumes that the bands of fcc, bcc, and hcp copper remain unchanged (or rigid) on alloying, so that the structural energy difference between any two lattices is given by

$$\Delta U = \Delta \left[ \int^{E_F} E n(E)\, dE \right],$$ 

(6.110)

where

$$N = \int^{E_F} n(E)\, dE.$$ 

(6.111)

It follows from eqn (6.109) that

$$\frac{d}{dN}(\Delta U) = \Delta \left[ \frac{dE_F}{dN} E_F n(E_F) \right] = \Delta E_F,$$ 

(6.112)

since on differentiating eqn (6.110) with respect to $N$ we have immediately

$$\frac{dE_F}{dN} n(E_F) = 1.$$ 

(6.113)

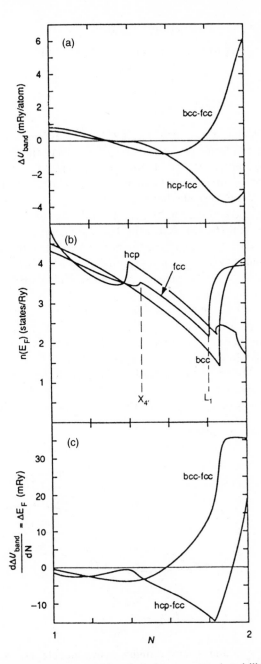

**Fig. 6.16** Analysis of fcc, bcc, and hcp relative structural stabilities within the rigid-band approximation for Cu–Zn alloys. (a) The difference in band energy as a function of band filling $N$ with respect to elemental rigid copper bands. (b) The density of states at the Fermi level $E_F$ for fcc, bcc, and hcp lattices as a function of band filling, $N$. (c) The difference in the Fermi energies $\Delta E_F$ as a function of band filling, $N$. (After Paxton *et al.* (1992).)

Further, it follows from eqs (6.111) and (6.112) that

$$\frac{d^2}{dN^2}(\Delta U) = \Delta\left[\frac{1}{n(E_F)}\right]. \tag{6.114}$$

Thus, the shape of the band energy difference curves in Fig. 6.16(a) can be understood in terms of the relative behaviour of the densities of states in the middle panel. In particular, from eqn (6.111), the stationary points in the upper curve correspond to band occupancies for which $\Delta E_F$ vanishes in panel (c). Moreover, whether the stationary point is a local maximum or minimum depends on the relative values of the density of states at the Fermi level through eqn (6.113). Thus, the bcc–fcc energy difference curve has a minimum around $N = 1.6$ where the bcc density of states is lowest, whereas the hcp–fcc curve has a minimum around $N = 1.9$, where the hcp density of states is lowest. The fcc structure is most stable around $N = 1$, where $\Delta E_F \approx 0$, and the fcc density of states is lowest.

We see that the structural trend from fcc → bcc → hcp is driven by the van Hove singularities in the densities of states. These arise whenever the band structure has zero slope as occurs at the bottom or top of the energy gaps at the Brillouin zone boundaries. The van Hove singularities at the bottom of the band gap at $X$ and at the top of the band gap at $L$ in fcc copper are marked $X_{4'}$ and $L_1$, respectively, in the middle panel of Fig. 6.16. It is, thus, not totally surprising that the reciprocal-space representation

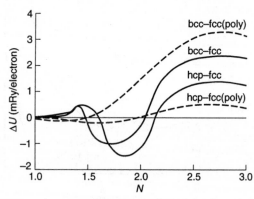

**Fig. 6.17** The structural-energy differences of a model Cu–Al alloy as a function of the band filling $N$, using an average Ashcroft empty-core pseudopotential with $R_c = 1.18$ au. The dashed curves correspond to the three-term analytic pair-potential approximation. The full curves correspond to the exact result that is obtained by correcting the difference between the Lindhard function and the rational polynomial approximation in Fig. 6.3 by a rapidly convergent summation over reciprocal space. (After Ward (1985).)

of second-order perturbation theory predicts energy difference curves in Fig. 6.17 that are very similar to those in the top panel of Fig. 6.16 (away from the copper-rich end where the *local* Ashcroft pseudopotential finds hcp the most stable structure). The strong curvature of the bcc–fcc and hcp–fcc curves can be reproduced as a function of band filling only by including explicitly the weak logarithmic singularity in the slope of the Lindhard response function at $q = 2k_F$. It is for this reason that these Hume–Rothery alloys are correctly termed *electron phases*, since this singularity is driven solely by the electron per atom ratio (through $2k_F$) and does not depend on the particular chemical constituents (through the pseudopotential). The actual prediction of the phase diagram in Fig. 6.15 requires a proper treatment of the total free energies of the different phases (see, for example, Turchi *et al.* (1991)). This takes us beyond the scope of the present chapter, but the interested reader can find the field reviewed in the excellent book by Ducastelle (1991) which brings together the recent developments in electron theory with those in statistical mechanics.

# References

Dagens, L., Rasolt, M., and Taylor, R. (1975). *Physical Review* **B11**, 2726.

Ducastelle, F. (1991). *Order and phase stability in alloys.* North-Holland, Amsterdam.

Finnis, M. W. (1974). *Journal of Physics* **F4**, 1645.

Friedel, J. (1952). *Philosophical Magazine* **43**, 153.

Gasiorowicz, S. (1974). *Quantum physics.* Wiley, New York.

Hafner, J. (1987). *From hamiltonians to phase diagrams: the electronic and statistical mechanical theory of sp-bonded metals and alloys.* Springer, Berlin.

Hafner, J. and Heine, V. (1983). *Journal of Physics* **F13**, 2473.

Heine, V. and Weaire, D. L. (1970). *Solid State Physics* **24**, 1.

Jones H. (1937). *Proceedings of the Physical Society* **A49**, 250.

Massalski, T. B. (Ed. in Chief). (1986). *Binary alloy phase diagrams* (ed. J. L. Murray, L. H. Bennett and H. Baker) Vol. 1. American Society for Metals, Metals Park, Ohio.

McMahan, A. K. and Moriarty, J. A. (1983). *Physical Review* **B27**, 3235.

Moriarty, J. A. (1983). *International Journal of Quantum Chemistry* **17**, 541.

Moriarty, J. A. (1988). *Physics Letters* **131**, 41.

Moriarty, J. A. and McMahan, A. K. (1982). *Physical Review Letters* **48**, 809.

Mott, N. F. and Jones, H. (1936). *Properties of metals and alloys*, chap. 7. Dover, New York.

Paxton, A. T., Methfessel, M. and Pettifor, D. G. (1992). Unpublished.

Pettifor, D. G. and Gelatt, C. D. (1983). *Atomistics of fracture* (ed R. M. Latanision and J. R. Pickens) p. 296. Plenum, New York.

Pettifor, D. G. and Ward, M. A. (1984). *Solid State Communications* **49**, 291.

Rasolt, M. and Taylor, R. (1975). *Physical Review* **B11**, 2717.

Taylor, R. (1978). *Journal of Physics* **F8**, 1699.

Turchi, P. E. A., Sluiter, M., Pinski, F. J., Johnson, D. D., Nicholson, D. M., Stocks, G. M., and Staunton, J. B. (1991). *Physical Review Letters* **67**, 1779.

Ward, M. A. (1985). *Analytic Pair Potentials for Simple Metals* (Ph.D. thesis). Imperial College of Science, Technology and Medicine, London.

Wyatt, T. K. (1991). *The Structural Stability of sp-bonded Metals.* (Third Year Undergraduate Project). Imperial College of Science, Technology and Medicine, London.

# 7
# Bonding of transition metals and semiconductors

## 7.1. Introduction

The only elements with cohesive energies greater than 7 eV per atom are sp-valent carbon (7.37 eV) and the sd-valent transition metals niobium (7.57 eV), tantalum (8.10 eV), tungsten (8.90 eV), rhenium (8.00 eV), and osmium (8.17 eV), as can be seen from Fig. 1.7. Interestingly, carbon with its *saturated* covalent $\sigma$ bonds does not have the highest cohesive energy, this honour belonging to the transition metal, tungsten. The transition metals are not describable by the conventional NFE model of the metallic bond since their valence d electrons remain relatively tightly bound to their parent atoms, forming *unsaturated* covalent bonds with their neighbours. These d bonds are responsible for the structural and cohesive properties of transition metals.

The bonding between d electrons in transition metals and sp electrons in semiconductors can thus be described within the same tight binding (TB) framework. We will begin this chapter, therefore, with an introduction to the TB prediction of energy band structure that extends our earlier linear combination of atomic orbitals (LCAO) treatment of molecules. Although the full band structure of transition metals requires the NFE sp band and its hybridization with the TB d band to be included, we will see, nevertheless, that the parabolic variation in the cohesive energy across the 4d or 5d transition-metal series is driven by the unsaturated bonding of the d electrons alone. Unlike the Pauling expression, eqn (3.34), the d bond contribution will be shown to drive not only the observed negative heats of formation but also the positive heats of formation that are found between certain pairs of transition metals. Finally, we will discuss the saturated covalent bond in sp-valent semiconductors and give an expression for the tetragonal shear constant that reflects the bond-bending resistance of the $sp^3$ hybrids.

## 7.2. The tight binding approximation

The application of the TB method to bulk systems is most easily introduced by first considering a lattice of atoms with overlapping s orbitals, $\psi_s$, and corresponding free atomic energy levels, $E_s$. Generalizing the LCAO method for the diatomic molecule to a periodic lattice of $\mathcal{N}$ atoms, we look for a crystal wave function, $\psi_{\mathbf{k}}$, that is a linear combination of the atomic orbitals, namely

$$\psi_{\mathbf{k}}(\mathbf{r}) = \mathcal{N}^{-1/2} \sum_{\mathbf{R}} e^{i\mathbf{k}\cdot\mathbf{R}} \psi_s(\mathbf{r} - \mathbf{R}). \tag{7.1}$$

The phase factor automatically guarantees that $\psi_{\mathbf{k}}(\mathbf{r})$ satisfies Bloch's theorem, eqn (5.30), since

$$\psi_{\mathbf{k}}(\mathbf{r} + \mathbf{S}) = e^{i\mathbf{k}\cdot\mathbf{S}} \mathcal{N}^{-1/2} \sum_{\mathbf{R}'=\mathbf{R}-\mathbf{S}} e^{i\mathbf{k}\cdot\mathbf{R}'} \psi_s(\mathbf{r} - \mathbf{R}') = e^{i\mathbf{k}\cdot\mathbf{S}} \psi_{\mathbf{k}}(\mathbf{r}). \tag{7.2}$$

The Schrödinger equation

$$\hat{H}\psi_{\mathbf{k}} = E_{\mathbf{k}}\psi_{\mathbf{k}} \tag{7.3}$$

then has the solution

$$E_{\mathbf{k}} = \frac{\int \psi_{\mathbf{k}}^* \hat{H} \psi_{\mathbf{k}} \, d\mathbf{r}}{\int \psi_{\mathbf{k}}^* \psi_{\mathbf{k}} \, d\mathbf{r}}. \tag{7.4}$$

Making the usual assumption that the crystalline potential, $V$, is given by the sum of overlapping atomic potentials, $v$, we have

$$\int \psi_{\mathbf{k}}^* \hat{H} \psi_{\mathbf{k}} \, d\mathbf{r} = \mathcal{N}^{-1} \sum_{\mathbf{R},\mathbf{S}} e^{i\mathbf{k}\cdot(\mathbf{R}-\mathbf{S})}$$

$$\times \int \psi_s(\mathbf{r} - \mathbf{S}) \left[ -\frac{\hbar^2}{2m} \nabla^2 + \sum_{\mathbf{T}} v(\mathbf{r} - \mathbf{T}) \right] \psi_s(\mathbf{r} - \mathbf{R}) \, d\mathbf{r} \tag{7.5}$$

and

$$\int \psi_{\mathbf{k}}^* \psi_{\mathbf{k}} \, d\mathbf{r} = \mathcal{N}^{-1} \sum_{\mathbf{R},\mathbf{S}} e^{i\mathbf{k}\cdot(\mathbf{R}-\mathbf{S})} \int \psi_s(\mathbf{r} - \mathbf{S}) \psi_s(\mathbf{r} - \mathbf{R}) \, d\mathbf{r}, \tag{7.6}$$

since $\psi_s^*(\mathbf{r}) = \psi_s(\mathbf{r})$ as $\psi_s$ is real. Neglecting the three centre integrals corresponding to $\mathbf{R} \neq \mathbf{S} \neq \mathbf{T}$ in eqn (7.5) and the overlap integrals, $\mathbf{R} \neq \mathbf{S}$, in eqn (7.6), we find the TB expression for the eigenvalue $E_{\mathbf{k}}$, namely

$$E_{\mathbf{k}} = E_s + \int \psi_s(\mathbf{r}) \left[ \sum_{\mathbf{R}}' v(\mathbf{r} - \mathbf{R}) \right] \psi_s(\mathbf{r}) \, d\mathbf{r} + \sum_{\mathbf{R}}' e^{i\mathbf{k}\cdot\mathbf{R}} \int \psi_s(\mathbf{r}) v(\mathbf{r}) \psi_s(\mathbf{r} - \mathbf{R}) \, d\mathbf{r}. \tag{7.7}$$

The second contribution on the right-hand side is the shift in the on-site energy due to the neighbouring atomic potentials. In the spirit of our earlier treatment of diatomic molecules we will neglect this crystal field term. It does not fundamentally alter the band structure of either transition metals or semiconductors. The band structure, $E(\mathbf{k})$, can, therefore, be written within the TB approximation as

$$E_{\mathbf{k}} = E_s + \sum_{\mathbf{R}}' e^{i\mathbf{k}\cdot\mathbf{R}} ss\sigma(R), \tag{7.8}$$

where $ss\sigma$ is the usual $\sigma$ bond integral between s orbitals.

The band structure for a simple cubic lattice may now be quickly found. Assuming that the bond integrals couple only to the six first nearest neighbours with position vectors, $\mathbf{R}$, equal to $(\pm a, 0, 0)$, $(0, \pm a, 0)$, and $(0, 0, \pm a)$, eqn (7.8) gives

$$E_{\mathbf{k}} = E_s + 2ss\sigma(\cos k_x a + \cos k_y a + \cos k_z a), \qquad (7.9)$$

where $\mathbf{k} = (k_x, k_y, k_z)$ and $ss\sigma = ss\sigma(R = a)$. Thus, the eigenvalues vary sinusoidally across the Brillouin zone. In particular, in the $|100\rangle$ and $|111\rangle$ directions we have

$$E_{\mathbf{k}} = E_s + 2ss\sigma \begin{cases} 2 + \cos ka & \text{for } \mathbf{k} = (k, 0, 0) \\ 3 \cos ka & \text{for } \mathbf{k} = (k, k, k). \end{cases} \qquad (7.10)$$

Therefore, as shown in Fig. 7.1(a), the bottom of the band is at the centre of the Brillouin zone $(0, 0, 0)$, whereas the top of the band is at the zone boundary $(\pi/a)(1, 1, 1)$, since $ss\sigma < 0$. It follows from eqn (7.1) that the bottom and top of the band correspond to perfect bonding and antibonding states, respectively, between all six neighbouring atoms, so that the width of the s band is $2|6ss\sigma|$, as expected. The corresponding density of states is shown in Fig. 7.1(b). The van Hove singularities, arising from the flat bands at the Brillouin zone boundaries, are clearly visible.

The structure of the TB p band may be obtained by writing $\psi_{\mathbf{k}}$ as a linear combination of the three p Bloch sums corresponding to the atomic $p_x$, $p_y$, and $p_z$ orbitals. That is,

$$\psi_{\mathbf{k}}(\mathbf{r}) = \mathcal{N}^{-1/2} \sum_{\alpha = x, y, z} c_\alpha \sum_{\mathbf{R}} e^{i\mathbf{k} \cdot \mathbf{R}} \psi_\alpha(\mathbf{r} - \mathbf{R}). \qquad (7.11)$$

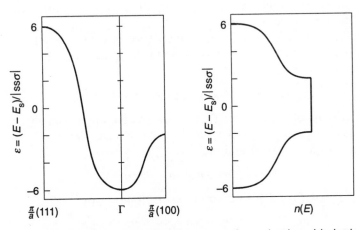

**Fig. 7.1** Left-hand panel: The s band structure for a simple cubic lattice in the $|100\rangle$ and $|111\rangle$ directions. Right-hand panel: The s band density of states for a simple cubic lattice.

Substituting eqn (7.11) into the Schrödinger eqn (7.3) leads to the $3 \times 3$ TB secular determinant for the p band, namely

$$|(E_p - E_k)\delta_{\alpha\alpha'} + T_{\alpha\alpha'}| = 0, \qquad (7.12)$$

where the matrix elements are given by

$$T_{\alpha\alpha'} = \sum_{\mathbf{R}}' e^{i\mathbf{k}\cdot\mathbf{R}} \int \psi_\alpha^*(\mathbf{r})v(\mathbf{r})\psi_{\alpha'}(\mathbf{r} - \mathbf{R})\, d\mathbf{r}. \qquad (7.13)$$

The bond integrals $\int \psi_\alpha^* v\psi_{\alpha'}\, d\mathbf{r}$ depend not only on the *distance*, $R$, but also on the *direction*, $\hat{\mathbf{R}} = (l, m, n)$, where $l$, $m$, and $n$ are the direction cosines, just as we have already found for the bent triatomic molecule $AH_2$, where from eqn (4.72)

$$\int \psi_s^* v\psi_x\, d\mathbf{r} = lsp\sigma. \qquad (7.14)$$

(The sign change between eqs (4.72) and (7.14) is due to the origin being located on the p orbital in Fig. 4.10 whereas above it is on the s orbital through eqn (7.13).) Similarly, as first shown by Slater and Koster in 1954,

$$\int \psi_x^* v\psi_x\, d\mathbf{r} = l^2 pp\sigma + (1 - l^2)pp\pi, \qquad (7.15)$$

$$\int \psi_x^* v\psi_y\, d\mathbf{r} = lmpp\sigma - lmpp\pi, \qquad (7.16)$$

and

$$\int \psi_x^* v\psi_z\, d\mathbf{r} = lnpp\sigma - lnpp\pi. \qquad (7.17)$$

All the other pp bond integrals, $\int \psi_\alpha^* v\psi_{\alpha'}\, d\mathbf{r}$ can be obtained from eqs (7.15)–(7.17) by cyclic permutation.

The band structure for a simple cubic lattice of p orbitals may now be found. Assuming only first nearest-neighbour hopping, the diagonal matrix elements are given by

$$T_{xx} = 2pp\sigma \cos k_x a + 2pp\pi(\cos k_y a + \cos k_z a) \qquad (7.18)$$

with $T_{yy}$ and $T_{zz}$ obtained from $T_{xx}$ by cyclic permutation. The off-diagonal matrix elements vanish for the simple cubic lattice. Hence in the $|100\rangle$ direction with $\mathbf{k} = (k, 0, 0)$ we have

$$E_\mathbf{k} = E_p + \begin{cases} 2pp\sigma \cos ka + 4pp\pi \\ 2pp\pi \cos ka + 2(pp\sigma + pp\pi) \\ 2pp\pi \cos ka + 2(pp\sigma + pp\pi). \end{cases} \qquad (7.19)$$

The ratio of the bond integrals for sp-valent elements was found by Harrison (1980) by fitting a nearest-neighbour TB model to the first principles band

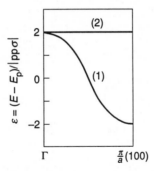

**Fig. 7.2** The p band structure for a simple cubic lattice in the $|100\rangle$ direction.

structure of *bulk* silicon and germanium. He obtained

$$\text{pp}\sigma:\text{pp}\pi:\text{sp}\sigma:\text{ss}\sigma = 2.31:-0.58:1.31:-1.00 \qquad (7.20)$$

Therefore, as a first approximation, we may neglect pp$\pi$ with respect to pp$\sigma$, and write

$$E_{\mathbf{k}} = E_{\mathrm{p}} + \begin{cases} 2\text{pp}\sigma \cos ka \\ 2\text{pp}\sigma \\ 2\text{pp}\sigma. \end{cases} \qquad (7.21)$$

This band structure is sketched in the $|100\rangle$ direction in Fig. 7.2. We see that, at the centre of the Brillouin zone, the eigenvalues are triply degenerate due to the cubic symmetry of the lattice. This degeneracy is partially lifted along the $|100\rangle$ symmetry direction. The singlet corresponds to bonding between the $\text{p}_x$ orbitals, whereas the doublet corresponds to the non-dispersive bonding between the neighbouring $\text{p}_y$ or $\text{p}_z$ orbitals. The degeneracy is totally lifted along a general $\mathbf{k}$ direction, as from eqs (7.12) and (7.13), there will then be three distinct eigenvalues.

Finally, the structure of the TB d band may be obtained by writing $\psi_{\mathbf{k}}$ as a linear combination of the five d Bloch sums corresponding to the five atomic orbitals illustrated in Fig. 2.15. This results in a $5 \times 5$ TB secular determinant,

$$|\mathscr{D} - EI| = 0, \qquad (7.22)$$

where

$$\mathscr{D}_{\alpha\alpha'} = E_{\mathrm{d}}\delta_{\alpha\alpha'} + \sum_{\mathbf{R}}' e^{i\mathbf{k}\cdot\mathbf{R}} \int \psi_{\alpha}^*(\mathbf{r})v(\mathbf{r})\psi_{\alpha'}(\mathbf{r} - \mathbf{R})\,d\mathbf{r} \qquad (7.23)$$

with $\alpha = xy,\ yz,\ xz,\ x^2 - y^2$, and $3z^2 - r^2$ respectively. The matrix elements can be expressed in terms of the three fundamental bond integrals dd$\sigma$, dd$\pi$, and dd$\delta$ by using Table I of Slater and Koster (1954). The ratios of these

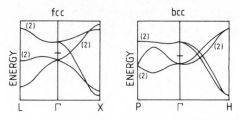

**Fig. 7.3** The fcc and bcc d band structure. (After Andersen (1973).).

bond integrals are given within the canonical band theory of Andersen (1973) by

$$dd\sigma : dd\pi : dd\delta = -6 : 4 : -1. \qquad (7.24)$$

Figure 7.3 shows the resulting band structure of the fcc and bcc lattices along the |111⟩ and |100⟩ directions in the Brillouin zone. We see that there are two energy levels at the centre of the Brillouin zone, $\Gamma$, one of which is triply degenerate, the other doubly degenerate. The former comprises the $T_{2g}$ orbitals $xy$, $yz$, and $xz$, which from Fig. 2.15 are equivalent to one another in a cubic environment. The latter comprises the $E_g$ orbitals $x^2 - y^2$ and $3z^2 - r^2$ which are not equivalent to the $T_{2g}$ orbitals because they point along the cubic axes. The degeneracy is partially lifted along the |111⟩ and |100⟩ symmetry directions as indicated in Fig. 7.3 because eigenfunctions which are equivalent at $\mathbf{k} = 0$ may become non-equivalent for $\mathbf{k} \neq 0$ due to the translational phase factor, $\exp(i\mathbf{k} \cdot \mathbf{R})$ (see Fig. 8.8 of Tinkham (1964)).

## 7.3. Hybrid NFE–TB bands

Transition metals are characterized by a fairly tightly bound d band of width $W$ that overlaps and hybridizes with a broader nearly-free-electron sp band as illustrated in Fig. 7.4. This difference in behaviour between the valence sp and d electrons arises from the d shell lying inside the outer valence s shell, thereby leading to small overlap between the d orbitals in the bulk.

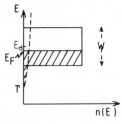

**Fig. 7.4** A schematic representation of transition metal sp (dashed curve) and d (solid curve) densities of states when sp-d hybridization is neglected.

For example, from eqn (2.55), the average radial distance of the hydrogenic 3d and 4s wave functions are in the ratio 0.44:1. Thus we expect the band structure of transition metals to be represented accurately by a hybrid NFE–TB secular equation of the form

$$\begin{vmatrix} \mathscr{C} - EI & \mathscr{H} \\ \mathscr{H}^\dagger & \mathscr{D} - EI \end{vmatrix} = 0, \tag{7.25}$$

where $\mathscr{C}$ and $\mathscr{D}$ are the sp-NFE and d-TB matrices respectively (cf eqs (5.38) and (7.22)). The hybridization matrix, $\mathscr{H}$, couples and mixes together the sp and d Bloch states that have the same symmetry.

A secular equation of this form can be derived directly from scattering theory, as first shown by Heine (1967), Hubbard (1967), and Jacobs (1968). They solved the Schrödinger equation (7.3) by regarding the lattice as a periodic array of scattering sites which individually scatter the electrons with a change in phase $\eta_l$ (see, for example, chapter 24 of Gasiorowicz (1974)). Transition metal sp-valence electrons are found to be scattered very weakly by the lattice, so that they exhibit NFE behaviour, with $\eta_0$ and $\eta_1$ close to zero. Transition-metal d electrons, on the other hand, are strongly scattered, the $l = 2$ phase shift exhibiting a resonance given by

$$\tan \eta_2(E) = \tfrac{1}{2}\Gamma/(E_d - E), \tag{7.26}$$

where $E_d$ and $\Gamma$ determine the position and width of the resonance, respectively. This resonant behaviour allows the scattering theory solution of the Schrödinger equation to be transformed exactly into the hybrid NFE–TB form. As a consequence, the numerous TB-bond integrals and hybridization-matrix elements are determined explicitly in terms of only the *two* resonant parameters, $E_d$ and $\Gamma$.

The band structure of nonmagnetic fcc and bcc iron is shown in Fig. 7.5, being computed from the hybrid NFE–TB secular equation with resonant parameters $E_d = 0.540$ Ry and $\Gamma = 0.088$ Ry. The NFE pseudopotential matrix elements were chosen by fitting the first principles band structure derived by Wood (1962) at the pure p states $N_{1'}$ ($v_{110} = 0.040$ Ry), $L_{2'}$ ($v_{111} = 0.039$ Ry), and $X_{4'}$ ($v_{200} = 0.034$ Ry). Comparing the band structure of iron in the $|100\rangle$ and $|111\rangle$ directions with the canonical d bands in Fig. 7.3, we see that there is only the *one* level with symmetry $\Delta_1$ and $\Lambda_1$, respectively, which hybridizes or mixes with the lowest NFE band, with the other four d levels of different symmetry unperturbed. Because of the canonical nature of the pure TB d bands reflected in eqn (7.24), the band structure of all the nonmagnetic fcc and bcc transition metals will be very similar to that shown in Fig. 7.5 for iron.

The transition-metal density of states $n(E)$ is not uniform throughout the band, as shown schematically in Fig. 7.4 but displays considerable structure that is characteristic of the given crystal lattice. This is seen in Fig. 7.6,

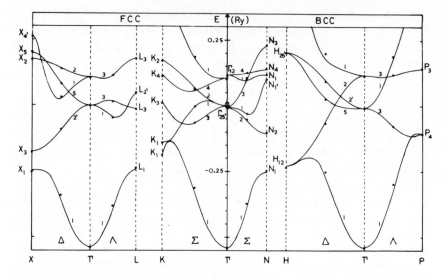

**Fig. 7.5** The hybrid NFE–TB band structure of fcc and bcc iron in the non-magnetic state. The solid circles represent the first principles energy levels of Wood (1962). (From Pettifor (1970*a*).)

which gives an early histogram of densities of states for bcc, fcc, and hcp lattices that were calculated from the hybrid NFE–TB secular equation with $E_d = 0.5$ Ry and $\Gamma = 0.06$ Ry. This structure in the densities of states is reflected in the observed behaviour of the electronic heat capacity across the nonmagnetic 4d and 5d transition-metal series. It follows from eqs (2.42), (2.43), and (2.44) that the electronic heat capacity varies linearly with temperature as

$$C_v = \gamma T, \tag{7.27}$$

where

$$\gamma = \tfrac{1}{3}\pi^2 k_B^2 n(E_F). \tag{7.28}$$

We see from Fig. 7.7 that the experimental values of $\gamma$ show the same trends across the series as that predicted by the model calculation. We will find later in subsequent sections that this structure in the densities of states is responsible for the ferromagnetism in bcc iron and the structural trend from hcp → bcc → hcp → fcc across the nonmagnetic 4d and 5d series.

## 7.4. The nature of the metallic bond in transition metals

The behaviour of the transition-metal bands as the atoms are brought together to form the solid may be evaluated within the Wigner–Seitz sphere approximation by imposing bonding, $R'_l = 0$, or antibonding, $R_l = 0$,

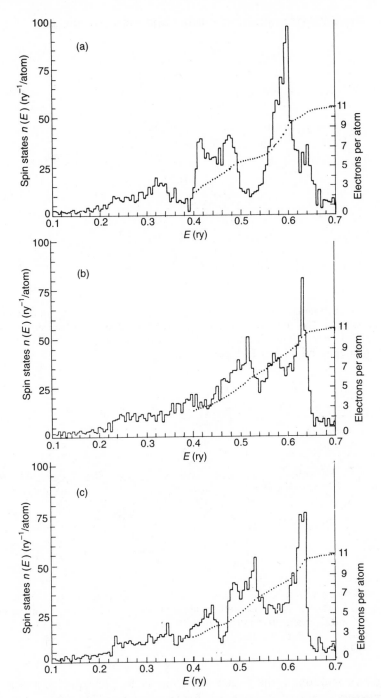

**Fig. 7.6** The densities of states for the three structures (a) bcc, (b) fcc, and (c) hcp for a model transition metal. The dotted curves represent the integrated density of states. (From Pettifor (1970*b*).)

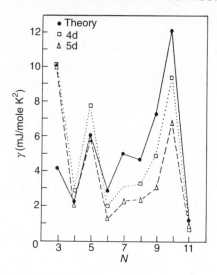

**Fig. 7.7** A comparison of the theoretical and experimental 4d and 5d electronic heat capacities. The theoretical values were obtained directly from eqn (7.28) and Fig. 7.6, neglecting any changes in the density of states due to bandwidth variation within the 4d and 5d series.

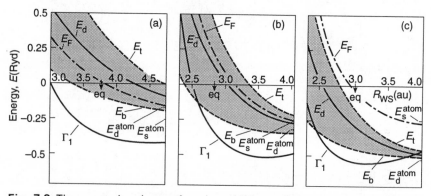

**Fig. 7.8** The energy bands as a function of Wigner–Seitz radius $R_{WS}$ for (a) Y, (b) Tc, and (c) Ag. The observed equilibrium Wigner–Seitz radii are marked eq. $E_d$, $E_t$, and $E_b$ mark the centre of gravity, and top and bottom of the d band respectively. (After Pettifor (1977).)

boundary conditions at the Wigner–Seitz radius, $R_{WS}$ (cf eqn (5.58)). Figure 7.8 shows the resultant energy bands for Y, Tc, and Ag that were computed by solving the local density approximation (LDA) Schrödinger equation self-consistently within the Wigner–Seitz sphere. We note six important influence the bonding and structural properties of transition metals:

1. We see that, whereas the 5s free-atom levels remains almost constant across the series, the 4d free-atom level drops from above the 5s level in Y to far below the 5s level in Ag, mirroring the behaviour already displayed in Fig. 2.17.

2. The bottom of the sp NFE conduction band, $\Gamma_1$, behaves in a similar fashion to that found by Wigner and Seitz in Fig. 5.14 for the alkali metal Na. Its volume dependence is also well described by a modified variant of the first term in eqn (5.63), namely

$$E_{\Gamma_1} = -\frac{3Z_c}{m_s R_{WS}} \left\{ 1 - \left( \frac{R_c}{R_{WS}} \right)^2 \right\}, \tag{7.29}$$

where $Z_c$ is the effective ionic charge which the sp conduction electrons see. The effective mass of the NFE bands, $m_s$ takes a value of approximately 0.8 across the 4d series. Table 7.1 shows that the fitted value of $Z_c$ is surprisingly close to unity. The fitted value of the Ashcroft empty-core radius, $R_c$, reflects the variation in the outer node of the free-atom 5s wave function, $R_{node}$,

$$R_c = 1.26 R_{node} \tag{7.30}$$

to within 1% across the series. This demonstrates unambiguously the importance of the repulsive core orthogonality constraints on the behaviour of the NFE sp band.

3. The centre of gravity of the d band $E_d$ moves up under compression as the electronic charge is confined into yet smaller Wigner–Seitz sphere volumes. The value of the potential, $v(R_{WS})$, at the Wigner–Seitz radius drops as approximately $-2Z_c/R_{WS}$ due to the exchange-correlation hole excluding one electron from the Wigner–Seitz sphere. The Hartree contribution to the LDA potential in eqn (2.61) is identically zero at the Wigner–Seitz radius because the Wigner–Seitz sphere is electrically neutral. It sets the energy zero in Fig. 7.8.

4. The volume dependence of the d band width is expected from a simple approximation within resonant scattering theory (Heine (1967)) to behave as

$$\Omega \frac{d}{d\Omega} (\ln W) = \frac{dW/d\Omega}{W/\Omega} \approx -\frac{2l+1}{3} = -\frac{5}{3}, \tag{7.31}$$

where the last equality follows for d electrons with $l = 2$. We see from Table 7.1 that this five-thirds relation is approximately satisfied across the series. The early transition elements have less tightly bound orbitals than the later transition elements, which is reflected in the magnitude of the logarithmic derivative increasing across the series from 3.86/3 for Y to 5.56/3 for Ag.

**Table 7.1** Band parameters for the 4d transition metals at their equilibrium atomic volumes. $W'$ is the logarithmic derivative of the bandwidth with respect to volume. (From Pettifor (1977).)

|  | $R_{WS}$ (au) | NFE sp band | | | TB d band | |
| --- | --- | --- | --- | --- | --- | --- |
|  |  | $Z_c$ | $R_c$(au) | $R_{node}$(au) | $W$(Ry) | $-3W'$ (eqn 7.31) |
| Y | 3.76 | 1.02 | 3.00 | 2.39 | 0.462 | 3.86 |
| Zr | 3.35 | 1.05 | 2.83 | 2.25 | 0.574 | 3.97 |
| Nb | 3.07 | 1.08 | 2.69 | 2.16 | 0.687 | 4.08 |
| Mo | 2.93 | 1.10 | 2.58 | 2.06 | 0.702 | 4.30 |
| Tc | 2.84 | 1.09 | 2.49 | 1.98 | 0.669 | 4.49 |
| Ru | 2.79 | 1.07 | 2.41 | 1.91 | 0.624 | 4.61 |
| Rh | 2.81 | 1.03 | 2.32 | 1.83 | 0.558 | 4.81 |
| Pd | 2.87 | 0.99 | 2.24 | 1.79 | 0.440 | 5.07 |
| Ag | 3.02 | 1.02 | 2.20 | 1.73 | 0.284 | 5.56 |

5. The relative occupancy of the NFE and TB bands varies under compression. We see from Fig. 7.9 that the d band occupancy $N_d$ increases dramatically for compressions about equilibrium for the early transition metals but that it holds steady for the later transition metals. For less than

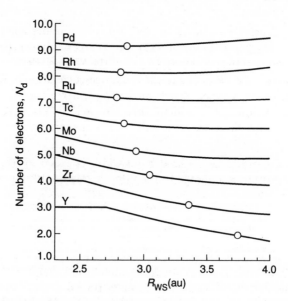

**Fig. 7.9** The d band occupancy $N_d$ as a function of the Wigner–Seitz radius, $R_{WS}$. The circles mark the equilibrium values.

half-full d bands, the Fermi energy moves *down* with respect to the centre of gravity of the d band as the d band widens, whereas the bottom of the conduction band moves *up* due to core orthogonality constraints as shown in Fig. 7.8. This causes NFE sp electrons to flow into the TB d band. We will find in the next chapter that this has important implications for structural transitions under pressure and for the observed structural trend at normal pressure across the lanthanides in Table 1.1.

6. The values of the band edges and d band centre at the observed equilibrium atomic volumes of the 4d transition metals are shown in Fig. 7.10. We see that the Fermi energy displays a maximum near the middle of the series. Even though the d band is filling as we move across to the right, its centre of gravity is falling sufficiently fast to cause the Fermi energy to drop from molybdenum to silver. This trend is very important for understanding catalytic behaviour, since the interaction of molecules such as CO with a metal surface will depend on how the metallic states near the Fermi energy line up with the molecular states *in vacuo* (see, for example, Hoffmann (1988)).

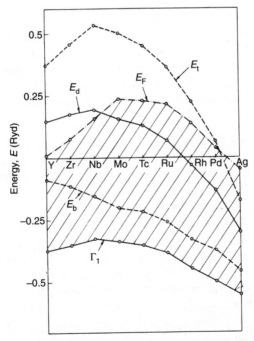

**Fig. 7.10** The variation of the energy bands across the 4d series at the equilibrium atomic volumes. Quantities $\Gamma_1$ and $E_d$ are the bottom of the sp band and centre of the d band respectively. Terms $E_F$, $E_t$, and $E_b$ represent the Fermi energy, and energy at the top and bottom of the d band respectively. (From Pettifor (1977).)

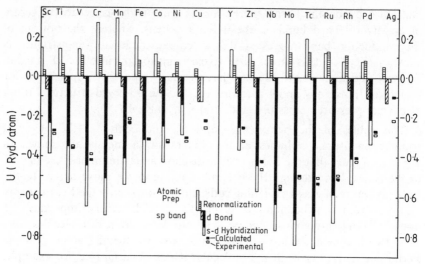

**Fig. 7.11** The contributions to the cohesive energy of the 3d and 4d transition metals. (After Gelatt *et al.* (1977).)

The binding energy of the transition metals may now be evaluated by summing over all the occupied states in the NFE and TB bands and subtracting off the large double-counting term which we mentioned earlier around eqn (6.54). Figure 7.11 shows the predicted cohesive energy across the 3d and 4d transition-metal series, which has been broken down into five physical contributions:

1. The *atomic preparation* energy is the energy required to take the observed ground state of the free atom and promote it into a singlet state with one valence s electron, this being the situation closest to the nonmagnetic bulk transition metal. We see that it is a positive contribution (or zero for the noble metals Cu and Ag). It is largest in the middle of the series where Hund's rule coupling leads to the special stability of the half-full d shell atoms.

2. The *renormalization* energy is the difference between the large repulsive contribution coming from the shift in the centre of gravity of the d band as the atoms bond together and the large double-counting term. The cancellation between these two terms is nearly complete leading to the small positive contribution of about one electronvolt that is shown in Fig. 7.11.

3. The *sp band* energy is the energy of the electrons in the NFE conduction band. It is small and negative at the ends of the series, decreasing to zero in the middle of the series due to the large core-orthogonality repulsion. Unlike the simple metals such as sodium, we see in Fig. 7.8 that the bottom of the transition metal band rises rapidly under compression due to the large fraction of the equilibrium atomic volume occupied by the core. The sp

electrons, therefore, exert a large outward pressure in transition metals at equilibrium.

4. The *d bond* energy is the energy of the d electrons that is measured with respect to the centre of gravity of the TB band, that is

$$U_{bond} = \int^{E_F} (E - E_d) n_d(E) \, dE \qquad (7.32)$$

where $n_d(E)$ is the TB d band density of states. It is analogous to the covalent bond energy defined earlier in eqn (4.30) for s-valent molecules. We see that it is large and negative in the middle of the series. Following Friedel (1969), and assuming a rectangular density of states of width, $W$, as illustrated in Fig. 7.4, the density of states per atom will take the constant value, $10/W$, because the d band must hold exactly ten electrons when it is full. The integral in eqn (7.32) may now be evaluated to give

$$U_{bond} = -\tfrac{1}{20} W N_d (10 - N_d) \qquad (7.33)$$

which displays the parabolic variation across the 3d and 4d series shown in Fig. 7.11. We see that it vanishes for the noble metals Cu and Ag with their nominally full valence d shells.

5. The *sp-d hybridization* energy is the contribution that results from turning on the hybridization matrix elements in eqn (7.25), resulting in the mixing between the NFE sp and TB d bands. As expected, it is negative, taking the approximately constant value of about 2 eV across the series.

## 7.5. The rectangular d band model of cohesion

The parabolic variation in the cohesive energy across the 4d series is driven by the d-bond contribution alone, as is clearly demonstrated by Fig. 7.11. The sizeable drop in the cohesive energy towards the middle of the 3d series is a free-atom phenomenon, resulting from the special stability of Cr and Mn atoms with their half-full d shells. The simplest model for describing the bonding of transition metals is, therefore, to write the binding energy per atom as

$$U = U_{rep} + U_{bond}, \qquad (7.34)$$

where, following eqn (4.29), the repulsive contribution is assumed to be pairwise, giving

$$U_{rep} = \frac{1}{2\mathcal{N}} \sum_{i,j}' \Phi_{rep}(R_{ij}). \qquad (7.35)$$

The d bond energy depends on the strength of the bond integrals dd$\sigma$, dd$\pi$, and dd$\delta$, which determine the TB density of states in eqn (7.32).

Within the rectangular d band model the bond energy is proportional to the bandwidth, $W$, through eqn (7.33). We may relate the bandwidth to the bond integrals by considering the second moment of the local density of states associated with atom i. Generalizing eqn (4.47) to the fact that we now have five d orbitals per site corresponding to $\alpha = xy$, $yz$, $xz$, $x^2 - y^2$ and $3z^2 - r^2$, we can write

$$\mu_2^{(i)} = \int_{-\infty}^{\infty} \varepsilon^2 n_d^{(i)}(\varepsilon) \, d\varepsilon = 2 \sum_j{}' \sum_{\alpha, \beta} H_{i\alpha, j\beta} H_{j\beta, i\alpha} \qquad (7.36)$$

where $\varepsilon = E - E_d$, and the prefactor, 2, accounts for the spin degeneracy. Giving the local density of states its constant value $10/W$ and realizing that in hopping from atom i to atom j the matrix, $H_{i\alpha, j\beta}$, is diagonal with elements $dd\sigma$, $dd\pi$, $dd\pi$, $dd\delta$, and $dd\delta$, if the $z$ axis is chosen along $\hat{\mathbf{R}}_{ij}$, we have for a lattice with coordination, $\varkappa$, that

$$\mu_2^{(i)} = \tfrac{10}{12}W^2 = 2\varkappa(5h^2), \qquad (7.37)$$
where

$$h^2 = \tfrac{1}{5}(dd\sigma^2 + 2dd\pi^2 + 2dd\delta^2). \qquad (7.38)$$

Thus, the bandwidth, $W$, can be written in terms of the root mean square bond integral, $h$, as

$$W = (12\varkappa)^{1/2}|h|, \qquad (7.39)$$

where we have chosen $h$ as a negative value like the dominant $dd\sigma$ contribution. Hence, the bond energy varies as the square root of the number of neighbours rather than linearly, thereby reflecting the *unsaturated* nature of the covalent bond in metals as discussed earlier in §5.7.

The cohesive energy, equilibrium atomic volume, and bulk modulus across a transition metal series may now be evaluated by choosing the following simple exponential forms for $\Phi(R)$ and $h(R)$, namely

$$\Phi(R) = aN_d^2 \, e^{-2\kappa R} \qquad (7.40)$$
and

$$h(R) = -bN_d \, e^{-\kappa R} \qquad (7.41)$$

where $a$ and $b$ are constants for a given series. The prefactors $N_d^2$ and $N_d$ are suggested by the respective dependence of $\Phi(R)$ and $h(R)$ on the atomic charge density. The explicit influence of the single valence s electron per atom is neglected. In the spirit of our earlier treatment of molecules, we have chosen $\Phi(R)$ proportional to $[h(R)]^2$, which corresponds to a degree of normalized hardness of the potential, $\alpha_h = \tfrac{1}{2}$. In practice, due to the large core-orthogonality repulsion, $\alpha_h$ is closer to two-thirds for transition metals. We will, however, retain the value, $\alpha_h = \tfrac{1}{2}$, in our following treatment of transition-metal cohesion and heats of formation, since it simplifies the algebra without affecting the basic underlying physical concepts. Substituting eqs (7.39)–(7.41) into eqs (7.35) and (7.33), the equilibrium expressions for the bandwidth, nearest-neighbour distance, cohesive energy, and bulk

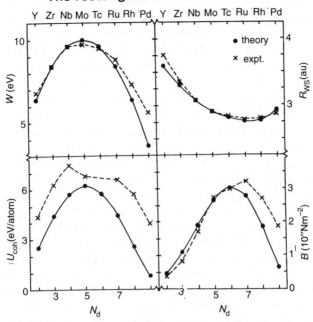

**Fig. 7.12** The theoretical (●) and experimental (×) values of the equilibrium band width, Wigner–Seitz radius, cohesive energy, and bulk modulus of the 4d transition metals. (From Pettifor (1987).)

modulus are given by

$$W = (3b^2/5a)N_d(10 - N_d) \tag{7.42}$$

$$R_0 = \kappa^{-1} \ln\{10a\sqrt{\varkappa}/[\sqrt{3}b(10 - N_d)]\} \tag{7.43}$$

$$U_{coh} = (3b^2/200a)[N_d(10 - N_d)]^2 \tag{7.44}$$

$$B = (2\sqrt{2}\kappa^2/9R_0)U_{coh}. \tag{7.45}$$

Figure 7.12 compares the theoretical predictions with the experimental values across the 4d series, assuming one valence s electron per atom and taking $\varkappa = 12$ corresponding to close-packed lattices. The 'experimental' values of the bandwidth are taken from the first principles LDA calculations in Table 7.1. The ratio $b^2/a$ is obtained by fitting a bandwidth of 10 eV for Mo with $N_d = 5$, so that from eqn (7.42) $b^2/a = \frac{2}{3}$ eV. The skewed parabolic behaviour of the observed equilibrium nearest-neighbour distance is found to be fitted by values of the inverse decay length $\kappa$ that vary linearly across the series as

$$\kappa = 0.435 + 0.075N_d. \tag{7.46}$$

This linear dependence is not unexpected from the linear variation in the free atomic d level that is observed across the 4d series in Fig. 2.17. The ratio, $a/b$, in eqn (7.44) is obtained by fitting the observed Wigner–Seitz radius of molybdenum, giving $a/b = 18.0$. It follows that $a = 216\,\text{eV}$ and $b = 12\,\text{eV}$ for the 4d series.

Thus, we see from Fig. 7.12 that the rectangular d band model is able to account qualitatively for the observed trends in cohesive energy, equilibrium nearest-neighbour distance, and bulk modulus across the nonmagnetic 4d (and 5d) transition-metal series. In particular, the parabolic behaviour of the cohesive energy reflects the initial filling of the bonding d states, followed by the antibonding states in Fig. 7.4. The skewed parabolic behaviour of the equilibrium nearest-neighbour distance, on the other hand, reflects the competition between the attractive bonding term, which varies parabolically with band filling, and the repulsive overlap term, which at a fixed internuclear separation decreases monotonically across the series as the size of the free atom contracts. We should note, however, that the d band model is a poor description of behaviour at the noble-metal end of the series, where the neglect of an explicit sp-d hybridization contribution leads to sizeable errors in the predicted cohesive energy and bulk modulus. Nevertheless, since the sp-d hybridization term is approximately constant across the 4d series at about 2 eV/atom as seen in Fig. 7.11, it may be neglected to a good approximation when predicting the trends in transition-metal heats of formation in the next section.

An interesting feature of this simple rectangular band model with $\alpha_h = \frac{1}{2}$ is that the cohesive energy in eqn (7.44) is independent of the coordination number, $z$, so that the diamond ($z = 4$), simple cubic ($z = 6$), and close-packed ($z = 12$), lattices, for example, would all be equally stable. The origin of this unexpected result may be traced back to the form of the binding energy, namely

$$U(R) = zC\,e^{-2\kappa R} - \sqrt{z}D\,e^{-\kappa R} \qquad (7.47)$$

where $C$ and $D$ are constants. The pairwise repulsive term is proportional to the number of nearest neighbours as expected, whereas the bonding term is proportional to the square root of the number of neighbours resulting from the unsaturated nature of the metallic bond. The equilibrium condition $U'(R = R_0) = 0$ gives immediately

$$e^{-\kappa R_0} = (D/2C)/\sqrt{z}, \qquad (7.48)$$

so that on substituting into eqn (7.47) the equilibrium binding energy is given by

$$U(R_0) = \tfrac{1}{4}D^2/C - \tfrac{1}{2}D^2/C = -\tfrac{1}{4}D^2/C, \qquad (7.49)$$

which is *independent* of the coordination of the lattice. This result is far from unsatisfactory, however, since we would not expect to make reliable structural predictions with a model in which the density of states has been smeared out into a uniform rectangular band, whose mean-square width is characterized by the second moment, $\mu_2$. Just as for the case of molecular structures, we will see in the next chapter that the structural trends within the transition metals are driven by higher moments, such as $\mu_3$ and $\mu_4$.

The equilibrium nearest-neighbour distance does depend on the coordination number, $\varkappa$, through either eqn (7.43) or eqn (7.48). We may write it in terms of the bond number, $n = N_d/\varkappa$, which gives the number of electrons per bond contributed by a given atom. The equilibrium nearest-neighbour distance then takes the identical form to that proposed by Pauling (1960) namely

$$R_n = R_1 - (1.15/\kappa) \log_{10} n, \tag{7.50}$$

where $R_1$ is the bond length associated with a single pair of electrons. For molybdenum, $\kappa = 0.81$ au$^{-1}$, so that the prefactor, $(1.15/\kappa)$, takes the value 0.75 Å, which compares with Pauling's empirical value of 0.6 Å for the metallic bond. The values of $\kappa$ that are used by the simple rectangular band model in eqn (7.46) lead, in fact, to a volume dependence of the bandwidth, $W$, that is in good agreement with the first principles LDA calculations in Table 7.1. For example, the model and first principles logarithmic derivatives for Y, Mo, and Rh are $(-1.3, -1.3)$, $(-1.4, -1.4)$, and $(-1.7, -1.6)$, respectively, which are close to the value of $-\frac{5}{3}$ predicted by Heine (1967); cf eqn (7.31). This demonstrates the fundamental correctness of this simple rectangular d band model for describing cohesion in transition metals.

## 7.6. The rectangular d band model of heats of formation

The heats of formation of equiatomic AB transition-metal alloys may be predicted by generalizing the rectangular d band model for the elements to the case of *disordered* binary systems, as illustrated in the lower panel of Fig. 7.13. Assuming that the A and B transition elements are characterized by bands of width $W_A$ and $W_B$, respectively, then they will mix together in the disordered AB alloy to create a common band with some new width, $W_{AB}$. The alloy bandwidth, $W_{AB}$ may be related to the elemental bond integrals, $h_{AA}$ and $h_{BB}$, and the atomic energy level mismatch, $\Delta E = E_B - E_A$, by evaluating the second moment of the total alloy density of states per atom $n_{AB}(E)$, namely

$$\mu_2^{AB} = \int_{-\infty}^{\infty} (E - \bar{E})^2 n_{AB}(E) \, dE \tag{7.51}$$

where $\bar{E} = \frac{1}{2}(E_A + E_B)$. Following a similar argument to the derivation of eqn (7.37) we find

$$\frac{1}{12}W_{AB}^2 = \frac{1}{2}\left\{\left[\frac{1}{2}\varkappa(h_{AA}^2 + h_{AB}^2) + \left(-\frac{\Delta E}{2}\right)^2\right] + \left[\frac{1}{2}\varkappa(h_{BA}^2 + h_{BB}^2) + \left(\frac{\Delta E}{2}\right)^2\right]\right\}. \tag{7.52}$$

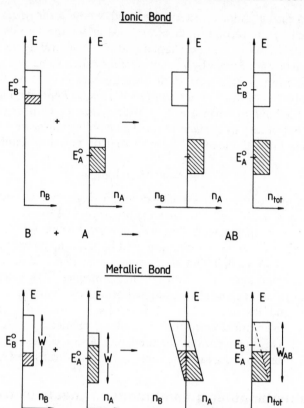

**Fig. 7.13** Schematic representation of the ionic rigid-band and metallic common-band models of bonding in binary AB systems. The quantities $E_A^0$ and $E_B^0$ give the free-atom energy levels, whereas the positions of $E_A$ and $E_B$ in the metallic bond reflect the small shift which takes place on alloy formation in order to maintain local charge neutrality.

The first term inside the first set of square brackets is due to the electron hopping between a central A atom and its $z$ neighbours, half of which are A and half of which are B on average in a disordered AB alloy. The second term inside the first set of square brackets arises from hopping twice on the same central atom A. The second set of square brackets is the corresponding central B atom contribution. The prefactor, $\frac{1}{2}$, outside the curly brackets results from the total alloy density of states being calculated per atom. We now make the realistic assumption that $h_{AB}$ is given by the geometric mean of $h_{AA}$ and $h_{BB}$, that is

$$h_{AB}(R) = [h_{AA}(R)h_{BB}(R)]^{1/2}. \qquad (7.53)$$

This is supported by the exponential form of the bond integrals in eqn (7.41).

Then

$$\tfrac{1}{12}W_{AB}^2 = z\bar{h}_{AB}^2 + \left(\frac{\Delta E}{2}\right)^2, \tag{7.54}$$

where $\bar{h}_{AB} = \tfrac{1}{2}(h_{AA} + h_{BB})$ is the average of the AA and BB bond integrals at the same first nearest-neighbour separation as the binary AB alloy. Thus, from eqn (7.39), we can write

$$W_{AB}^2 = \bar{W}_{AB}^2 + 3(\Delta E)^2, \tag{7.55}$$

where $\bar{W}_{AB} = \tfrac{1}{2}(W_A^{AB} + W_B^{AB})$ is the average bandwidth of the elemental transition metals at the *same* atomic volume as that of the AB alloy. Equation (7.55) generalizes expression (3.20) for the AB molecule to the bulk. Hence,

$$W_{AB} = (1 + 3\delta^2)^{1/2}\bar{W}_{AB} \tag{7.56}$$

where $\delta = \Delta E/\bar{W}_{AB}$ is the normalized atomic energy-level mismatch.

This common band model of the binary alloy is very different from the ionic model that is illustrated in the upper panel of Fig. 7.13. In the ionic picture the density of states of the binary alloy is assumed to be a rigid superposition of the elemental densities of states, so that charge flows from the B site to the A site in order to equilibrate the Fermi energy, thereby setting up an ionic bond through the electrostatic Madelung interaction. In the common band picture, on the other hand, the local density of states on a given atom changes when the atom is taken from its elemental environment to that of the alloy. The band in the alloy is formed by the quantum mechanical overlap of neighbouring atomic wave functions, so that all the local densities of states have a common bandwidth. Moreover, we know from moment theory, eqn (4.49), that the centre of gravity of the local density of states must coincide with the local on-site energy level. Thus, we may represent the local densities of states, $n_A(E)$ and $n_B(E)$, by skewed rectangular bands as shown in the lower panel of Fig. 7.13. If the skew rectangular density of states $n_A(E)$ takes the values $10(1 + \alpha)/W_{AB}$ and $10(1 - \alpha)/W_{AB}$ at the bottom and top of the band respectively, then for the centre of gravity to coincide with $E_A$, we must have that the degree of skewing is given by

$$\alpha = 3\Delta E/W_{AB}. \tag{7.57}$$

The charge transfer, $-Q_d e$, which accompanies a given atomic energy-level separation, $\Delta E$, in the binary AB alloy may be obtained by filling the local densities of states up to the Fermi level as shown in Fig. 7.13. For the skew rectangular local densities of states this gives

$$Q_d = \tfrac{1}{2}\Delta N + \tfrac{3}{10}\bar{N}(10 - \bar{N})\Delta E/W_{AB}, \tag{7.58}$$

where $\bar{N} = \tfrac{1}{2}(N_d^A + N_d^B)$ and $\Delta N = N_d^B - N_d^A$. The positions and labelling of the bands in Fig. 7.13 correspond to $\Delta E > 0$ and $\Delta N < 0$. Thus, the first term represents the charge flowing from the high-valence A atom to the low-valence B atom, whereas the second term represents charge flowing in the opposite direction from the higher to lower atomic energy level.

We have seen in the previous chapter that metals exhibit perfect screening. Therefore, the atomic energy-level mismatch will adjust itself in the metallic alloy to maintain *local charge neutrality* (LCN). This is a problem of self-consistency in that the alloy bandwidth, $W_{AB}$, on the right-hand side of eqn (7.58) depends on the atomic energy-level mismatch, $\Delta E$, through eqn (7.56). Solving both equations with $Q_d = 0$ yields the value of the normalized atomic energy-level mismatch for local charge neutrality, namely

$$\delta_{LCN} = -\{\tfrac{9}{25}[\bar{N}(10 - \bar{N})]^2 - 3(\Delta N)^2\}^{-1/2}\Delta N. \tag{7.59}$$

Thus, for bands that are nearly half-full with $\bar{N} \approx 5$, we have that the atomic energy-level mismatch for LCN will be given by

$$\Delta E_{LCN} \approx -\tfrac{1}{15}\bar{W}_{AB}\Delta N \approx -\tfrac{2}{3}\Delta N \text{ eV}, \tag{7.60}$$

where we have chosen a value of 10 eV for the average bandwidth. This is to be compared with the difference in the d energy levels of *free* transition metal atoms of about $-1$ eV per valence difference across the 4d series (cf Fig. 2.17).

The binding energy per atom of the AB alloy may, therefore, be written very simply as

$$U^{AB} = U_{rep}^{AB} + U_{bond}^{AB}, \tag{7.61}$$

since no ionic Madelung term appears as a result of local charge neutrality. If the alloy is completely *disordered*, then the repulsive energy per atom is given by

$$U_{rep}^{AB} = \tfrac{1}{2}z[\tfrac{1}{2}(\Phi_{AA} + \Phi_{AB}) + \tfrac{1}{2}(\Phi_{BA} + \Phi_{BB})]/2 \tag{7.62}$$

where $z$ is the coordination. The first term inside the square brackets corresponds to the repulsion between the central A atom and its $z$ neighbours, half of which are A and half of which are B, on average. The second term inside the square brackets is the corresponding central B atom contribution. The quantity inside the square brackets is divided by two because we are considering energies per atom. Assuming that $\Phi_{AB}$ is given by the geometric mean of $\Phi_{AA}$ and $\Phi_{BB}$ as suggested by the exponential form of the repulsive pair potential, we have from eqs (7.40) and (7.41) that

$$\Phi_{\alpha\beta}(R) = (a/b^2)h_{\alpha\beta}^2(R), \tag{7.63}$$

where $\alpha$ and $\beta$ can take either the values A or B. Hence, the repulsive energy per atom of a disordered binary AB alloy with $z$ nearest neighbours a distance, $R$, apart can be written

$$U_{rep}^{AB} = \tfrac{1}{2}(a/b^2)z\bar{h}_{AB}^2 = \tfrac{1}{24}(a/b^2)\bar{W}_{AB}^2. \tag{7.64}$$

The d bond energy for the disordered AB alloy is defined by

$$U_{\text{bond}}^{AB} = \frac{1}{2}\left[\int^{E_F} (E - E_A)n_A(E)\,dE + \int^{E_F} (E - E_B)n_B(E)\,dE\right]. \quad (7.65)$$

It may be written in terms of the *total* density of states, $n_{AB}(E)$, as

$$U_{\text{bond}}^{AB} = \int^{E_F} E n_{AB}(E)\,dE - \tfrac{1}{2}(N_A E_A + N_B E_B). \quad (7.66)$$

Hence, within the rectangular d band model for the AB alloy density of states, from eqs (7.33) the bond energy becomes

$$U_{\text{bond}}^{AB}(\Delta E) = -\tfrac{1}{20}W_{AB}\bar{N}(10 - \bar{N}) - \tfrac{1}{4}\Delta N \Delta E \quad (7.67)$$

This expression is stationary with respect to small variations in $\Delta E$ for just that value of $\Delta E = \Delta E_{LCN}$, which results from filling up the skew rectangular *partial* densities of states and requiring local charge neutrality. Thus, this simple model is internally consistent.

The bond energy per atom, when the local charge neutrality condition eqn (7.59) is satisfied, is given by

$$U_{\text{bond}}^{AB} = f_0(\bar{N}, \Delta N)U_{\text{bond}}(\bar{N}), \quad (7.68)$$

where

$$f_0(\bar{N}, \Delta N) = \{1 - \tfrac{25}{3}(\Delta N)^2/[\bar{N}(10 - \bar{N})]^2\}^{1/2} \quad (7.69)$$

and

$$U_{\text{bond}}(\bar{N}) = -\tfrac{1}{20}\bar{W}_{AB}\bar{N}(10 - \bar{N}). \quad (7.70)$$

The bond energy, $U_{\text{bond}}(\bar{N})$, is just that of the AB alloy within the *virtual crystal approximation* (VCA) in which the electrons see only the average potential $\bar{v} = \tfrac{1}{2}(v_A + v_B)$ at each site, so that all atoms would be characterized by the same atomic energy level $\bar{E} = \tfrac{1}{2}(E_A + E_B)$. The prefactor $f_0(\bar{N}, \Delta N)$ represents the loss of bond energy with respect to this average VCA state resulting from the actual mismatch in the atomic energy levels on the A and B sites, $\Delta E_{AB}$.

The heat of formation $\Delta H$ may now be found by comparing the binding energy of the AB alloy at its equilibrium nearest-neighbour separation, $R_0^{AB}$, with that of the A and B elemental transition metals at their equilibrium nearest neighbour distances, $R_0^A$ and $R_0^B$, respectively, as shown in Fig. 7.14. We may use the structural energy difference theorem of §4.3 to write down this small energy difference directly as

$$\Delta H = U_{\text{bond}}^{AB}(\hat{R}_0^{AB}) - \tfrac{1}{2}[U_{\text{bond}}^A(R_0^A) + U_{\text{bond}}^B(R_0^B)], \quad (7.71)$$

where $\hat{R}_0^{AB}$ is the nearest-neighbour separation in the AB alloy, such that the alloy and the elemental metals display the same repulsive energy, that is

$$U_{\text{rep}}^{AB}(\hat{R}_0^{AB}) = \tfrac{1}{2}[U_{\text{rep}}^A(R_0^A) + U_{\text{rep}}^B(R_0^B)]. \quad (7.72)$$

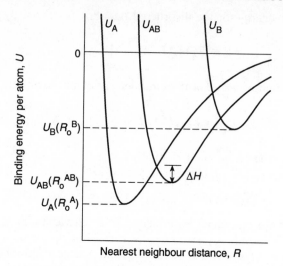

**Fig. 7.14** The binding energy curves for the elemental A and B transition metals and the binary AB alloy. The heat of formation is given by $\Delta H = U_{AB} - \frac{1}{2}(U_A + U_B)$, where the binding energies are evaluated at the appropriate equilibrium positions as shown.

This repulsive energy constraint implies from eqn (7.64) that the bond energy for the AB alloy must be evaluated for a value of

$$\bar{W}_{AB} = [\tfrac{1}{2}(W_A^2 + W_B^2)]^{1/2}. \qquad (7.73)$$

Substituting eqs (7.68) and (7.33) into eqn (7.71) and using eqn (7.42) for the elemental bandwidths, the heat of formation can be expressed as the sum of two terms, namely

$$\Delta H = \Delta H_{\text{integral}}^{\text{bond}} + \Delta H_{\text{order}}^{\text{bond}}, \qquad (7.74)$$

where

$$\Delta H_{\text{integral}}^{\text{bond}} = (3b^2/200a)(5 - \bar{N})^2(\Delta N)^2 \qquad (7.75)$$

and

$$\Delta H_{\text{order}}^{\text{bond}} = -(b^2/400a)[3\bar{N}(10 - \bar{N}) - 50](\Delta N)^2. \qquad (7.76)$$

Fourth-order contributions and higher have been neglected. The first contribution gives the change in the elemental bond energies as their *bond integrals* are changed to take the elemental bandwidths, $W_A$ and $W_B$, to the new common bandwidth, $\bar{W}_{AB}$, that is

$$\Delta H_{\text{integral}}^{\text{bond}} = -\tfrac{1}{2}[\tfrac{1}{20}(\bar{W}_{AB} - W_A)N_A(10 - N_A) + \tfrac{1}{20}(\bar{W}_{AB} - W_B)N_B(10 - N_B)]. \qquad (7.77)$$

The second contribution gives the change in the bond energy as the elemental bands of width, $\bar{W}_{AB}$ mix together to form the new common band in the

AB alloy. Since the AA and BB bond integrals have already been adjusted to give identical elemental bandwidths, this contribution reflects from eqn (4.62) the change in *bond order* that accompanies alloy formation. It is given by

$$\Delta H_{\text{order}}^{\text{bond}} = U_{\text{bond}}^{\text{AB}} - \tfrac{1}{2}[-\tfrac{1}{20}\bar{W}_{\text{AB}}N_{\text{A}}(10 - N_{\text{A}}) - \tfrac{1}{20}\bar{W}_{\text{AB}}N_{\text{B}}(10 - N_{\text{B}})], \quad (7.78)$$

which to second order in $\Delta N$ can be written

$$\Delta H_{\text{order}}^{\text{bond}} = \{-\tfrac{1}{80} + \tfrac{5}{24}[\bar{N}(10 - \bar{N})]^{-1}\}\bar{W}_{\text{AB}}(\Delta N)^2. \quad (7.79)$$

The first contribution inside the curly brackets represents the change in the bond energy within the virtual crystal approximation, that is

$$\Delta H_{\text{VCA}} = -\tfrac{1}{20}\bar{W}_{\text{AB}}\{\bar{N}(10 - \bar{N}) - \tfrac{1}{2}[N_{\text{A}}(10 - N_{\text{A}}) + N_{\text{B}}(10 - N_{\text{B}})]\}. \quad (7.80)$$

The second contribution inside the curly brackets of eqn (7.79) represents the loss of bonding due to the atomic energy-level mismatch in the alloy. Since from eqs (7.73) and (7.42) $\bar{W}_{\text{AB}} = (3b^2/5a)\bar{N}(10 - \bar{N})$, to the zeroth order in $\Delta N$, eqs (7.79) and (7.76) are identical to second order.

Figure 7.15 shows the different contributions to the normalized heats of formation $\Delta H/(\Delta N)^2$ as a function of the average d band filling, $\bar{N}$, for the case of 4d transition-metal alloys, where $b^2/a = \tfrac{2}{3}$ eV. We see that, whereas the dashed VCA curve is always negative, the dotted bond order curve is negative only for average d band occupancies, $2 \leq \bar{N} \leq 8$, because otherwise the loss in energy due to the atomic energy-level mismatch drives this contribution positive. The bond integral contribution due to differences in the elemental bandwidths is always positive. The resultant total heat of formation is negative for average band fillings, such that

$$3\tfrac{1}{3} < \bar{N} < 6\tfrac{2}{3}, \quad (7.81)$$

as can be proved by setting $\Delta H$ in eqn (7.74) equal to zero. The prediction of positive heats of formation within the TB model accounts for the bare patches that are observed amongst the transition-metal compounds within the AB structure map at the end of the book. The experimental values for NbMo and RhPd, and the values predicted by Miedema's semi-empirical scheme for disordered 4d transition-metal alloys with $\Delta N = 1$ or 2, have also been plotted in Fig. 7.15. Larger values of $\Delta N$ have been excluded for comparison, since terms beyond second order have been neglected in eqs (7.75) and (7.76).

We see that the simple rectangular d band model reproduces the behaviour found by experiment and predicted by Miedema's semi-empirical scheme. However, we must stress that the TB model does not give credence to any theory that bases the heat of formation of transition-metal alloys on *ionic* Madelung contributions that arise from electronegativity differences between the constituent atoms because in the metallic state the atoms are perfectly screened and, hence, locally charge neutral. Instead, the TB model supports

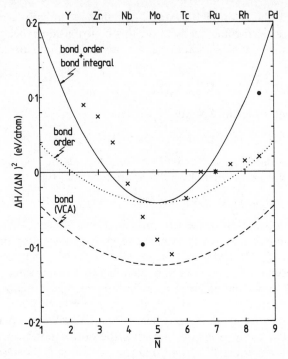

**Fig. 7.15** The contributions to the normalized heats of formation, $\Delta H/(\Delta N)^2$, for the case of 4d transition metal alloys. The experimental and semi-empirical values of Miedema *et al.* (1980) for $\Delta N = 1$ and 2 are given by the solid dots and crosses respectively. (From Pettifor (1987).)

the earlier suggestion by Brewer in 1968 that the most stable transition-metal alloys would comprise elements from groups at the opposite ends of the transition-metal series, such as Y and Pd. These groups have very few bonding electrons, since they have nearly empty or full d shells. Mixing these elements together results in a dramatic increase in their bond order as the electrons would be shared in the bonding states of the alloy corresponding to a half-full band, thereby leading to a sizeable lowering of their *covalent* bond energy.

## 7.7. The saturated bond in semiconductors

The energy gap in tetrahedral semiconductors, such as carbon, silicon, and germanium, is neither a consequence of the long-range periodicity of the lattice (as in the $\mu \ll 1$ regime of the Kronig–Penney model in Fig. 5.6) nor a result of the atomic energy levels not yet having broadened enough to form a continuous band (as in the $\mu \gg 1$ regime of the Kronig–Penney model

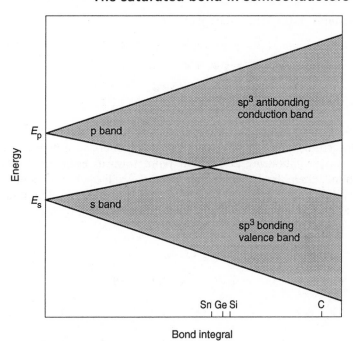

**Fig. 7.16** Schematic illustration of the opening-up of the hybridization band gap in the energy bands of tetrahedral sp-valent solids once the strength of the bond integral becomes sufficiently large. The positions of C, Si, Ge, and Sn along the horizontal axis are marked according to the relative values of their experimental band gaps. (After Cox (1991).)

in Fig 5.6). It is instead a *hybridization gap* that opens up within the very broad band of $sp^3$ states as illustrated schematically in Fig. 7.16. The free atom sp splittings in C, Si, Ge, and Sn fall within the range $7.5 \pm 1$ eV, so that we can regard $\Delta E_{sp} = E_p - E_s$ as being approximately constant within these tetrahedral semiconductors. However, the bond integrals increase by about a factor of two in going from tin to carbon as a result of the decreasing nearest-neighbour bond length (cf Fig. 3.13 for the dimers, $C_2$ and $Si_2$). Figure 7.16 is, thus, drawn approximately to scale for silicon with an sp splitting of 7 eV, a band gap of 1.1 eV, and a total bandwidth of about 20 eV. Elements, C, Ge, and Sn have been marked on this schematic diagram to correspond to band gaps of 5.5, 0.7, and 0.1 eV, respectively.

The origin of this hybridization gap in tetrahedral semiconductors can best be understood by taking the four $sp^3$ hybrid orbitals as our starting basis rather than the four free atomic orbitals s, $p_x$, $p_y$, and $p_z$. As is well known (see, for example, McWeeny (1979)), the former are linear combinations of the latter, such that

$$\left.\begin{array}{l} \phi_1 = \tfrac{1}{2}(\psi_s + \psi_x + \psi_y + \psi_z) \\ \phi_2 = \tfrac{1}{2}(\psi_s + \psi_x - \psi_y + \psi_z) \\ \phi_3 = \tfrac{1}{2}(\psi_s - \psi_x + \psi_y - \psi_z) \\ \phi_4 = \tfrac{1}{2}(\psi_s - \psi_x - \psi_y + \psi_z) \end{array}\right\}. \qquad (7.82)$$

These four orbitals point out towards the four first nearest neighbours along the tetrahedral bonds. If we consider only the interaction between the two hybrids on neighbouring atoms that point towards each other along the same bond, then the hybrid energy level $E_0$ will split into bonding and anti-bonding levels separated by $2|h|$ as shown in Fig. 7.17. It follows from eqn (7.82) that the energy of the hybrid orbital is given by

$$E_0 = \tfrac{1}{4}(E_s + 3E_p), \qquad (7.83)$$

and the hybrid bond integral is given by,

$$h = \tfrac{1}{4}(\mathrm{ss}\sigma - 2\sqrt{3}\mathrm{sp}\sigma - 3\mathrm{pp}\sigma). \qquad (7.84)$$

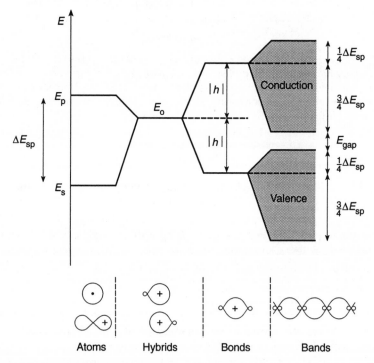

**Fig. 7.17** Successive transformations of linear combinations of atomic orbitals, beginning with atomic s and p orbitals, proceeding to $sp^3$ hybrids, forming bond orbitals and antibond orbitals, then coupling to form the valence band and conduction band respectively. (After Harrison (1980).).

These two sharp bonding and antibonding levels are then broadened into a band of states by coupling between different hybrids on the *same* atom, since

$$\int \phi_\alpha \hat{H} \phi_\beta \, d\mathbf{r} = -\tfrac{1}{4}(E_\mathrm{p} - E_\mathrm{s}) = -\tfrac{1}{4}\Delta E_\mathrm{sp}, \quad \text{for } \alpha \neq \beta, \qquad (7.85)$$

which we have seen in Fig. 7.16 is a sizeable matrix element for the tetrahedral semiconductors. This coupling allows an electron in a given bonding state between atoms $i$ and $j$ to hop via atom $i$ into a neighbouring bonding state between atoms $i$ and $k$, and thence on through the lattice. Writing the bond orbital between atoms $i$ and $j$ as

$$\psi_{ij}^+ = \frac{1}{\sqrt{2}}(\phi_{ij}^{(i)} + \phi_{ij}^{(j)}), \qquad (7.86)$$

where $\phi_{ij}^{(i)}$ and $\phi_{ij}^{(j)}$ are the appropriate hybrids on atoms, $i$ and $j$ which point along the bond $ij$, then we can evaluate the coupling between the *bond orbitals* on neighbouring bonds as

$$\int \psi_{ij}^+ \hat{H} \psi_{ik}^+ \, d\mathbf{r} = \frac{1}{2} \int (\phi_{ij}^{(i)} + \phi_{ij}^{(j)}) \hat{H} (\phi_{ik}^{(i)} + \phi_{ik}^{(k)}) \, d\mathbf{r}. \qquad (7.87)$$

Since we are neglecting overlap between hybrids on different atoms unless they point along the same bond, only one contribution remains in eqn (7.87), so that

$$\int \psi_{ij}^+ \hat{H} \psi_{ik}^+ \, d\mathbf{r} = \frac{1}{2} \int \phi_{ij}^{(i)} \hat{H} \phi_{ik}^{(i)} \, d\mathbf{r} = -\tfrac{1}{8}\Delta E_\mathrm{sp} \qquad (7.88)$$

from eqn (7.85).

The energy of the bottom of the valence band $E_\mathrm{b}^\mathrm{v}$ may now be found, since it corresponds to the most bonding state illustrated in the top panel of Fig. 7.18, in which each bond orbital is in phase with the six neighbouring bond orbitals. Hence, its energy is given by

$$E_\mathrm{b}^\mathrm{v} = E_0 + h + 6(-\tfrac{1}{8}\Delta E_\mathrm{sp}) = E_0 + h - \tfrac{3}{4}\Delta E_\mathrm{sp}. \qquad (7.89)$$

Similarly, the energy of the top of the valence band, $E_\mathrm{t}^\mathrm{v}$, may be found, since it corresponds to the most antibonding combination of the bond orbitals in which neighbouring bond orbitals are 180° out of phase, as shown in the lower panel of Fig. 7.18. Hence, its energy is given by

$$E_\mathrm{t}^\mathrm{v} = E_0 + h + (2 - 4)(-\tfrac{1}{8}\Delta E_\mathrm{sp}) = E_0 + h + \tfrac{1}{4}\Delta E_\mathrm{sp}. \qquad (7.90)$$

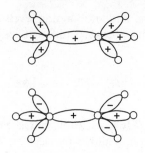

**Fig. 7.18** The *upper figure* illustrates the most bonding state of the valence band in which all bond orbitals between neighbouring pairs of atoms are in phase. The *lower figure* illustrates the most antibonding state of the valence band in which the bond orbitals between neighbouring pairs of atoms are 180° out of the phase. (After Heine (1971).)

Thus we find a valence band of width

$$E_t^v - E_b^v = \Delta E_{sp}, \tag{7.91}$$

as sketched in Fig. 7.17. Similarly, by considering the antibond orbitals, $\psi_{ij}^- = (1/\sqrt{2})(\phi_{ij}^{(i)} - \phi_{ij}^{(j)})$, the conduction band may be shown to broaden in a similar fashion. Hence, a hybridization gap opens up in the $sp^3$ band provided

$$\Delta E_{sp}/2|h| < 1. \tag{7.92}$$

We see from Fig. 7.16 that this condition is only just satisfied for tetrahedral tin.

This description of the opening-up of the hybridization gap has neglected any coupling between the bond orbitals, $\psi_{ij}^+$, and the antibond orbitals, $\psi_{ij}^-$, so that the bond would remain totally *saturated* with a bond order from eqn (4.67) of

$$\Theta_{ij} = \tfrac{1}{2}(N_{ij}^+ - N_{ij}^-) = \tfrac{1}{2}(2 - 0) = 1. \tag{7.93}$$

In practice, just as we have already found for the isolated sp-valent dimer in §3.7, there will be mixing between the bond and antibond orbitals that becomes increasingly marked as the sp splitting, $\Delta E_{sp}$, increases from zero. The influence of both the sp splitting and the local atomic environment on the bond order may be estimated by taking the model density of states shown in Fig. 7.19. The left-hand panel illustrates schematically the delta-function density of states associated with the bond and antibond orbitals, $\psi_{ij}^+$ and $\psi_{ij}^-$, respectively for the isolated sp-valent dimer with $\Delta E_{sp} = 0$. The right-hand panel shows that mixing between the bond and antibond orbitals occurs as the atomic energy-level splitting, $\Delta E_{sp}$, is turned on and the dimer

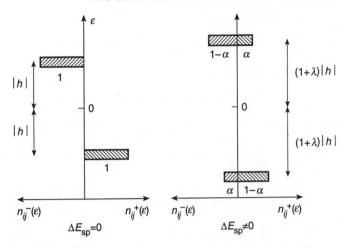

**Fig. 7.19** Model densities of states representing the mixing between the bond orbitals, $\psi_{ij}^+$, and antibond orbitals, $\psi_{ij}^-$, when $\Delta E_{sp} \neq 0$. The shaded rectangles represent delta functions with weights given by the attached numbers. The left-hand panel corresponds to $\Delta E_{sp} = 0$ when there is no mixing between the bond orbitals with energy $-|h|$ and the antibond orbitals with energy $+|h|$. The right-hand panel corresponds to $\Delta E_{sp} \neq 0$ when the mixing between the states pushes them apart by the factor $(1 + \lambda)$ and reduces their original weights by the factor $(1 - \alpha)$.

is embedded in its bulk environment. This mixing or hybridization causes the energy levels to move apart as shown by the factor $(1 + \lambda)$ and the weight in the bond orbital to be reduced in the lower level by the factor $(1 - \alpha)$ and to be increased in the upper level by $\alpha$, so that the corresponding density of states per spin is given by

$$n_{ij}^+(\varepsilon) = (1 - \alpha)\delta[\varepsilon + (1 + \lambda)|h|] + \alpha\delta[\varepsilon - (1 + \lambda)|h|], \qquad (7.94)$$

where $\varepsilon = E - E_0$. Similarly, the density of states per spin associated with the antibond orbital can be written

$$n_{ij}^-(\varepsilon) = \alpha\delta[\varepsilon + (1 + \lambda)|h|] + (1 - \alpha)\delta[\varepsilon - (1 + \lambda)|h|]. \qquad (7.95)$$

The mixing parameters, $\alpha$ and $\lambda$, may be obtained from the first and second moments of the densities of states. The first moment of $n_{ij}^+(\varepsilon)$ gives the bond orbital energy with respect to $E_0$ of $-|h|$, so that from eqn (7.94)

$$\mu_1^+ = -(1 - 2\alpha)(1 + \lambda)|h| = -|h|. \qquad (7.96)$$

Hence

$$1 - 2\alpha = (1 + \lambda)^{-1}. \qquad (7.97)$$

Similarly, the second moment of the total density of states per orbital,

$n_{ij} = \frac{1}{2}(n_{ij}^+ + n_{ij}^-)$, is given by

$$\mu_2 = \frac{1}{2}(\mu_2^+ + \mu_2^-) = (1 + \lambda)^2 |h|^2, \qquad (7.98)$$

so that substituting into eqn (7.97) we have

$$1 - 2\alpha = |h|/\mu_2^{1/2}. \qquad (7.99)$$

But from eqs (7.94) and (7.95), the bond order for a half-full band is given by

$$\Theta_{ij} = 1 - 2\alpha, \qquad (7.100)$$

so that from eqn (7.99)

$$\Theta_{ij} = |h(R_{ij})|/\mu_2^{1/2}. \qquad (7.101)$$

The second moment $\mu_2$ can be written explicitly in terms of the atomic energy-level splitting $\Delta E_{sp}$ and any interaction between the bond ij and its local atomic environment. It follows from eqs (7.86) and (7.98) that

$$\mu_2 = \frac{1}{2}(\mu_2^{(i)} + \mu_2^{(j)}), \qquad (7.102)$$

where $\mu_2^{(i)}$ and $\mu_2^{(j)}$ are the second moment of the local density of states associated with the hybrid $\phi_{ij}^{(i)}$ and $\phi_{ij}^{(j)}$ on atoms i and j respectively. By summing over all paths of length two that start and end on these two orbitals we find

$$\mu_2 = \frac{3}{16}(\Delta E_{sp})^2 + h^2(R_{ij}) + \sum_{k \neq i,j} \frac{1}{2}[h^2(R_{ik})g(\theta_{jik}) + h^2(R_{jk})g(\theta_{ijk})], \qquad (7.103)$$

where the angular factor is

$$g(\theta) = \frac{11}{32} + \frac{3}{8}\cos\theta + \frac{9}{32}\cos 2\theta. \qquad (7.104)$$

The first term in eqn (7.103) corresponds to hopping twice on the same atom, so that from eqn (7.82) it takes the value,

$$\frac{1}{4}(E_s^2 + 3E_p^2) = \frac{1}{4}[(-\frac{3}{4}\Delta E_{sp})^2 + 3(\frac{1}{4}\Delta E_{sp})^2], \qquad (7.105)$$

since the zero of energy has been taken as $E_0 = \frac{1}{4}(E_s + 3E_p)$. The second term in eqn (7.103) corresponds to hopping back and forth along the bond ij. The third term in eqn (7.100) corresponds to hopping from atoms i or j to a neighbouring atom k, and back again. The angular factor, $g(\theta)$, follows from the angular dependence of the bond integrals that is given in eqs (7.14)–(7.17). It is plotted in Fig. 7.20, where we see that both $g(\theta)$ and $g'(\theta)$ vanish at the tetrahedral angle, $\theta_0 = \cos^{-1}(-\frac{1}{3})$.

The bond order for tetrahedral *elemental* semiconductors at equilibrium may, therefore, be written as

$$\Theta = [1 + \frac{3}{4}(\Delta E_{sp}/2h)^2]^{-1/2}. \qquad (7.106)$$

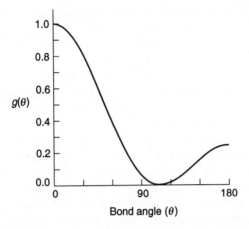

**Fig. 7.20** The angular factor $g(\theta)$ as a function of the bond angle $\theta$. Note that both $g$ and $g'$ vanish at the tetrahedral angle $\theta_0 = \cos^{-1}(-\frac{1}{3})$.

Consequently, the bond is fully saturated for $\Delta E_{sp} = 0$ with a bond order of 1, but it is only partially saturated by the time the gap closes for $\Delta E_{sp}/2|h| = 1$ (cf eqn (7.92)) when the bond order equals 0.76. This simple second moment model has been extended to include the *compound* semiconductors. The resultant values of the bond order are given in Table 7.2. We see that the bonds in tetrahedral carbon and silicon are almost fully saturated, but those in zinc selenide and cadmium telluride are only about 75% saturated due partly to the mismatch in the sp orbitals between chemically distinct atoms.

The binding energy per atom of an elemental semiconductor can be written as

$$U = \frac{1}{2\mathcal{N}} \sum_{i,j}' [\Phi(R_{ij}) + 2h(R_{ij})\Theta_{ij}] + U_{prom}, \tag{7.107}$$

**Table 7.2** The bond order for tetrahedral semiconductors. (From Alinaghian *et al.* 1994).)

| Elemental | | Compound | |
|---|---|---|---|
| C | 0.94 | AlP | 0.83 |
| Si | 0.86 | GaAs | 0.79 |
| Ge | 0.81 | InSb | 0.77 |
| Sn | 0.79 | ZnSe | 0.75 |
| | | CdTe | 0.73 |

where we have used eqn (4.62) to write the bond energy as the product of the bond integral and the bond order. The energy, $U_{prom}$ is required to promote the free-atom $s^2p^2$ state into the state appropriate for the bulk. For the hybrid $sp^3$ state this energy is $\Delta E_{sp} = E_p - E_s$, which is large and positive, taking the value of 8.5 eV for carbon. The promotion energy is responsible for the fact that even though diamond has saturated sp-valent bonds, it has less cohesive energy than unsaturated sd-valent niobium, which requires only about 1 eV of the promotion energy (cf Fig. 7.11).

The expression for the binding energy, eqn (7.107), is not a simple sum over pair potentials, since, three-body interactions enter through the environment dependence of the bond order that is given by eqs (7.101) and (7.103). The bond angle-dependent function, $g(\theta)$, in fact, plays a crucial role in determining the stiffness of the tetrahedral bonds to bending. For example, under tetragonal shear the second-order change in energy may be evaluated analytically (Alinaghian *et al.* (1994).) Neglecting any second-order changes in the promotion energy, the tetragonal shear constant is given by

$$C' = (3\sqrt{3}|h_0|/4R_0^3)\Theta_0^3 g''(\theta_0), \qquad (7.108)$$

where $h_0$, $\Theta_0$, $R_0$, and $\theta_0$ are the equilibrium values of the bond integral, bond order, bond length, and bond angle respectively, and $g''(\theta_0)$ is the curvature of the angular function, which takes the value unity, from eqn (7.104), since $\cos \theta_0 = -\frac{1}{3}$. Thus, as is well known, it is the angular character of the bonding that stabilizes the tetrahedral lattice against shear.

An expression of the type (7.101), which gives the bond order explicitly in terms of the positions of the neighbouring atoms, is called a bond order potential (BOP). Angularly dependent bond order potentials were first derived *heuristically* for the elemental semiconductors by Tersoff (1988). We will see in the next chapter that a many-body expansion for the bond order may be derived *exactly* within the TB model.

# References

Alinaghian, P., Nishitani, S. R., and Pettifor, D. G. (1994). *Philosophical Magazine* **B69**, 889.
Andersen, O. K. (1973). *Solid State Communications* **13**, 133.
Brewer, L. (1968). *Science* **161**, 115.
Cox, P. A. (1991). *The electronic structure and chemistry of solids*. Oxford University Press.
Friedel, J. (1969). *The physics of metals* (ed. J. M. Ziman), p. 494. Cambridge University Press, New York.
Gasiorowicz, S. (1974). *Quantum physics*. New York.
Gelatt, C. D., Ehrenreich, H., and Watson, R. E. (1977). *Physical Review* **B15**, 1613.
Harrison, W. A. (1980). *Electronic structure and the properties of solids*. Freeman, San Francisco.
Heine, V. (1967). *Physical Review* **153**, 673.

Heine, V. (1971). *Journal of Physics* **C4**, L221.

Hoffmann, R. (1988). *Solids and surfaces: a chemist's view of bonding in extended structures.* VCH, New York.

Hubbard, J. (1967). *Proceedings of the Physical Society* **92**, 921.

Jacobs, R. L. (1968). *Journal of Physics* **C1**, 492.

McWeeny, R. (1979). *Coulson's valence.* Oxford University Press.

Miedema, A. R., Chatel, P. F., and de Boer, F. R. (1980). *Physica* **B100**, 1.

Pauling, L. (1960). *The nature of the chemical bond.* Cornell University Press, Ithaca, New York State.

Pettifor, D. G. (1970*a*). *Physical Review* **B2**, 3031.

Pettifor, D. G. (1970*b*). *Journal of Physics* **C3**, 367.

Pettifor, D. G. (1977). *Journal of Physics* **F7**, 613.

Pettifor, D. G. (1987). *Solid State Physics* **40**, 43.

Slater, J. C. and Koster, G. F. (1954). *Physical Review* **94**, 1498.

Tersoff, J. (1988). *Physical Review* **B38**, 9902.

Tinkham, M. (1964). *Group theory and quantum mechanics.* McGraw-Hill, New York.

Wood, J. H. (1962). *Physical Review* **126**, 517.

# 8

# Structural trends within solids

## 8.1. Introduction

Elemental solids and binary compounds display numerous well-defined structural trends. We saw in chapter 1 that we would like theory to be able to explain at least the following:

(1) the change from close-packed structures on the left-hand side of the periodic table to the more open structures on the right-hand side;
(2) the 8-N rule which gives the number of neighbours expected for $N \geq 4$, and exceptions to the rule such as graphitic carbon and dimeric nitrogen and oxygen;
(3) the structural trend from hcp $\rightarrow$ bcc $\rightarrow$ hcp $\rightarrow$ fcc across the transition-metal series, and exceptions to the trend such as manganese and iron;
(4) the structural trend from the La structure type to the Sm structure type to hcp across the lanthanides; and
(5) the structural trends within the AB structure map between NaCl, CsCl, CrB, FeB, FeSi, NiAs, and MnP structure types that are taken by pd-bonded binary compounds.

In this chapter we will show that the tight binding (TB) description of the covalent bond is able to provide a simple and unifying explanation for the above structural trends and behaviour. We will see that the ideas already introduced in chapter 4 on the structures of small molecules may be taken over to these infinite bulk systems. In particular, we will find that the trends in structural stability across the periodic table or within the structure maps can be linked directly to the topology of the local atomic environment through the moments theorem of Ducastelle and Cyrot-Lackmann (1971).

## 8.2. Saturated versus unsaturated bonds

The 8-N rule states that the number of bonds (or local coordination, $\varkappa$) equals 8 minus the number of the periodic group. This rule is illustrated in Fig. 1.2 where we see that for $N = 7$ the halogens take dimeric structure types with $\varkappa = 1$, for $N = 6$ the chalcogenides selenium and tellurium take helical chain structures with $\varkappa = 2$, for $N = 5$ the pnictides arsenic, antimony, and bismuth take a puckered layer structure with $\varkappa = 3$, and for $N = 4$ the semiconductors

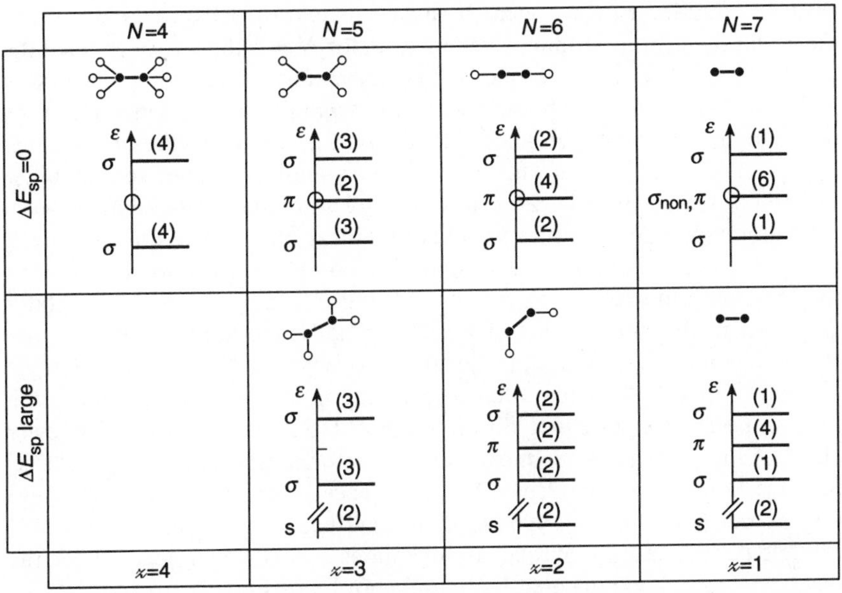

| | $N=4$ | $N=5$ | $N=6$ | $N=7$ |
|---|---|---|---|---|
| $\Delta E_{sp}=0$ | $\varepsilon$ $\sigma$ (4) $\sigma$ (4) | $\varepsilon$ $\sigma$ (3) $\pi$ (2) $\sigma$ (3) | $\varepsilon$ $\sigma$ (2) $\pi$ (4) $\sigma$ (2) | $\varepsilon$ $\sigma$ (1) $\sigma_{non},\pi$ (6) $\sigma$ (1) |
| $\Delta E_{sp}$ large | | $\varepsilon$ $\sigma$ (3) $\sigma$ (3) s (2) | $\varepsilon$ $\sigma$ (2) $\pi$ (2) $\sigma$ (2) s (2) | $\varepsilon$ $\sigma$ (1) $\pi$ (4) $\sigma$ (1) s (2) |
| | $\varkappa=4$ | $\varkappa=3$ | $\varkappa=2$ | $\varkappa=1$ |

**Fig. 8.1** The eigenspectra of diamond ($\varkappa = 4$), graphite or arsenic ($\varkappa = 3$), linear or helical chain ($\varkappa = 2$), and dimer ($\varkappa = 1$) for the case of zero sp splitting (upper panel) and large sp splitting (lower panel). The $\pi$ bonding is assumed to vanish (i.e. pp$\pi = 0$) and the interaction between hybrids on different atoms that do not point along the same bond has been neglected.

silicon and germanium take the tetrahedral diamond structure with $\varkappa = 4$. This 8-N rule is usually rationalized within a valence bond framework by assuming that single saturated covalent bonds are formed with neighbours, thereby completing the stable octet shell of electrons about each sp-valent atom.

A molecular orbital (MO) or tight binding (TB) description *appears* to provide a similar conclusion. For simplicity let us consider the case, pp$\pi = 0$, and let us neglect any interaction between hybrids on different atoms that do not point along the same bond. The upper panel in Fig. 8.1 shows the resultant energy levels for zero sp splitting, that is, $\Delta E_{sp} = E_p - E_s = 0$. The $N = 4$ panel gives the splitting of the sp$^3$ hybrids into the bonding and antibonding levels that we found earlier in Fig. 7.17. These two energy levels remain sharp because there is no coupling of the neighbouring bond orbitals or antibond orbitals together, since $\Delta E_{sp} = 0$. Thus, these levels are four-fold degenerate, corresponding to the four uncoupled neighbouring bonds about a given atom. With four valence electrons per atom the bonding level will be fully occupied, so that there will be a single saturated covalent $\sigma$ bond between every neighbour in the tetrahedral solid.

The tetrahedral crystal structure is assumed to change if another valence electron is added to the system because this extra electron would go into the destabilizing antibonding level. Instead, for $N = 5$ the lattice chooses the graphitic structure because it can then form fully saturated sp$^2$ hybrid $\sigma$ bonds with its three neighbours with the remaining two valence electrons per atom going into the nonbonding $\pi$ states, as shown in the upper panel of Fig. 8.1. The addition of a further electron is similarly argued to destabilize the graphitic structure because it would go into the antibonding $\sigma$ level. Instead, for $N = 6$ the lattice chooses the linear chain structure, because it can then form fully saturated sp hybrid $\sigma$ bonds with its two neighbours, the remaining four electrons per atom going into the nonbonding $\pi$ states as shown in the upper panel of Fig. 8.1. Finally, the addition of a further electron would destabilize the linear chain so that $N = 7$ takes dimeric structure types with their six-fold degenerate nonbonding level comprising the non-bonding $\sigma$ state in addition to the $\pi$ states (cf eqn (3.62)). The lower panel of Fig. 8.1 shows the most stable structures expected if $\Delta E_{sp}$ is large; then the bonding will be driven by the valence p orbitals alone, so that we may expect bond angles of 90° to be formed between the three orthogonal orbitals p$_x$, p$_y$, and p$_z$. Similar arguments as for $\Delta E_{sp} = 0$ account for the pnictides taking the puckered three-fold coordinated layer structure (cf Fig. 1.5) and the chalcogenides taking the helical two-fold coordinated chain structure (cf Fig. 1.2) as shown in the lower panel of Fig. 8.1.

The above argument that sp-valent solids will take saturated single bonds according to the 8-N rule is, however, flawed. The alert reader will already have noticed that there is *no* difference in energy between the two competing structure types for a given $N$ with the energy levels as drawn in Fig. 8.1. Putting an extra electron in the antibonding level cancels the presence of one electron in the bonding level, so that it is exactly equivalent to putting two electrons into the nonbonding level of the neighbouring structure type. This is seen explicitly in the left-hand panel of Fig. 8.2 where for $N = 5$ the three-fold and four-fold coordinated structure types have the same bond energy, for $N = 6$ the two-fold, three-fold, and four-fold coordinated structure types have the same bond energy, and for $N = 7$ all four structure types have the same bond energy. Thus, no structural differentiation can be made for $N \geq 5$.

It might be argued that the structural trend across the top panel of Fig. 8.1 is driven by the increasing strength of the hybrid bond integral as the percentage of s character changes from sp$^3 \to$ sp$^2 \to$ sp. If we choose the z-axis along the axis of the bond ij, then the bonding hybrids may be written as

$$\phi_\lambda^{(i)} = \frac{1}{\sqrt{1 + \lambda^2}} (\psi_{is} + \lambda \psi_{iz}) \tag{8.1}$$

and

$$\phi_\lambda^{(j)} = \frac{1}{\sqrt{1 + \lambda^2}} (\psi_{js} - \lambda \psi_{jz}) \tag{8.2}$$

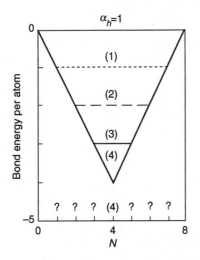
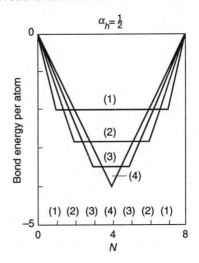

**Fig. 8.2** The bond energy per atom of the four-fold, three-fold, two-fold and one-fold coordinated lattices as a function of the number of valence electrons per atom, $N$, for the cases of degrees of normalized hardness of the potential, $\alpha_h = 1$ (left-hand panel) and $\alpha_h = \frac{1}{2}$ (right-hand panel). The hard core potential of the left-hand panel cannot differentiate between different structure types for $N \geq 5$, whereas the realistic $\alpha_h = \frac{1}{2}$ potential gives the structural sequence for $4\text{-} \to 3\text{-} \to 2\text{-} \to 1$-fold coordination of the (8-N) rule.

where the fraction of s admixture in the hybrid is $1/(1 + \lambda^2)$. Thus, sp, sp$^2$, and sp$^3$ hybrids correspond to $\lambda = 1$, $\sqrt{2}$, and $\sqrt{3}$ respectively. The bond integral between these directed hybrids then takes the value,

$$h_\lambda = \int \phi_\lambda^{(i)} \hat{H} \phi_\lambda^{(j)} \, d\mathbf{r} = \frac{\text{ss}\sigma - 2\lambda\text{sp}\sigma - \lambda^2\text{pp}\sigma}{1 + \lambda^2}. \tag{8.3}$$

This is plotted in Fig. 8.3 as a function of the percentage of s character, $100/(1 + \lambda^2)$, for two different choices of the ratios pp$\sigma$:sp$\sigma$:ss$\sigma$, namely the solid curve where their magnitudes are all equal (cf eqn (3.58)), and the dashed curve where their magnitudes are set according to Harrison's (1980) prescription for the bulk band structure (cf eqn (7.20)). The full curve indeed predicts that $|h_{\text{sp}}| > |h_{\text{sp}^2}| > |h_{\text{sp}^3}|$, which would drive the structural trend across the top panel of Fig. 8.1. Unfortunately, the more realistic dashed curve provides no such justification for the (8-N) rule, since $|h_{\text{sp}}| < |h_{\text{sp}^2}| \approx |h_{\text{sp}^3}|$.

In fact, the origin of the (8-N) rule resides in the delicate balance between the repulsive overlap forces and the attractive covalent bond forces. The bond lengths are not invariant as drawn in Fig. 8.1, since the atoms do not behave as hard spheres with fixed nearest-neighbour distances. Assuming a repulsive pair potential

$$\Phi(R) = A[h(R)]^2, \tag{8.4}$$

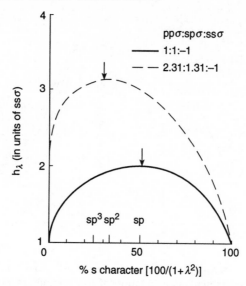

**Fig. 8.3** The bond integral, $h_\lambda$, between directed hybrids as a function of the percentage of s character of the hybrid, $100/(1 + \lambda^2)$, for two choices of the ratios $pp\sigma:sp\sigma:ss\sigma$.

which is a good approximation for the sp-valent elements (cf eqn (3.47)), then the difference in the equilibrium binding energies between the two structure types is given by the structural energy difference theorem as

$$\Delta U = [\Delta U_{\text{bond}}]_{\Delta\mu_2 = 0} \tag{8.5}$$

(cf eqn (4.54)). That is, the bond energies must be compared only once the bond lengths have been adjusted, so that all the eigenspectra have the same second moment or mean-square width. Taking the tetrahedral bond integral $h_4$ as reference, the other bond integrals $h_1$, $h_2$, and $h_3$ for the dimer, linear chain, and graphitic sheet will satisfy

$$h_1^2 = 2h_2^2 = 3h_3^2 = 4h_4^2. \tag{8.6}$$

The resultant bond energies are shown in the right-hand panel of Fig. 8.2, where we see the trend from 4-fold → 3-fold → 2-fold → 1-fold coordinated structure types as $N$ changes from $4 \to 5 \to 6 \to 7$ respectively.

The (8-N) rule is driven by the fact that, although the bond energy initially remains the same in Fig. 8.1 on going from the diamond to the graphitic structure for $N = 5$, or from the graphitic structure to the linear chain for $N = 6$, or from the linear chain to the dimer for $N = 7$, the repulsive energy decreases due to the lower coordination number. This causes the bond lengths to shrink, thereby providing the additional cohesion that is reflected in the right-hand panel of Fig. 8.2. Of course, *after* the bond lengths have equilibrated in their new local environment, we would then have from eqn (8.6) that

$$|h_{sp}| > |h_{sp^2}| > |h_{sp^3}| \tag{8.7}$$

as found experimentally.

There are many exceptions to the (8-N) rule: for $N = 4$ carbon takes the graphitic ground state structure with $\varkappa = 3$ and lead takes the fcc structure with $\varkappa = 12$; for $N = 5$ nitrogen is dimeric with $\varkappa = 2$; and for $N = 6$ oxygen is dimeric with $\varkappa = 2$ and polonium is simple cubic with $\varkappa = 6$. It might have been hoped that the simple analytic expression for the bond order, eqn (7.101), might have provided some structural differentiation for $N = 4$ between the diamond lattice with its *saturated* bonds and either the simple cubic, simple hexagonal, or close-packed lattices with their *unsaturated* bonds. In fact, it predicts that they have identical equilibrium binding energies (Nishitani *et al.* (1994)).

This totally unexpected result can be proved as follows. Within a first nearest-neighbour model, the bond order of these lattices may be written from eqs (7.101) and (7.103) as

$$\Theta_{ij} = \left[ \sum_{k=1}^{\varkappa} g(\theta_{jik}) \right]^{-1/2}, \tag{8.8}$$

where $k$ sums over the $\varkappa$ nearest neighbours to atom, i, $\theta_{jik}$ is the corresponding bond angle, and $g(\theta)$ is the bond angle dependent function,

$$g(\theta) = \tfrac{11}{32} + \tfrac{3}{8} \cos \theta + \tfrac{9}{32}(2 \cos^2 \theta - 1), \tag{8.9}$$

as $\cos 2\theta = 2 \cos^2 \theta - 1$. Summing the angular function over the nearest neighbours yields

$$\sum_{k=1}^{\varkappa} g(\theta_{jik}) = \tfrac{11}{32}\varkappa + 0 + \tfrac{9}{32}[2(\tfrac{1}{3}) - 1]\varkappa \tag{8.10}$$

so that

$$\Theta = 2/\sqrt{\varkappa} \tag{8.11}$$

for these three-dimensional lattices. The bond orders of the diamond ($\varkappa = 4$), simple cubic ($\varkappa = 6$), simple hexagonal ($\varkappa = 8$), and close-packed ($\varkappa = 12$) lattices are predicted to take the values 1.00, 0.82, 0.71, and 0.58 respectively. Thus, the bonds in the open diamond lattice are saturated, whereas those in the close-packed lattices are unsaturated, as expected. Nevertheless, since there are $\varkappa$ bonds, the total bond energy per atom will vary as $\sqrt{\varkappa}$, so that from §7.5 the cohesive energy will be independent of the local coordination (with the realistic choice of the degree of normalized hardness of the potential, $\alpha_h = \tfrac{1}{2}$).

Thus, there is no *a priori* reason for saturated bonds to form a more stable structure than unsaturated bonds. Although the individual bond energy is larger, there are fewer neighbours in the open structures with saturated bonds. It is, therefore, not surprising that there are many exceptions to the 8-N rule. In the next section we will show how the structural trends within the sp-valent elements can be rationalized within a simple TB model.

## 8.3. Structural trends within the sp-valent elements

We have seen in the previous chapter that the total binding energy per atom of an elemental sp-valent system may be written within the TB approximation as the sum of three terms, namely

$$U = U_{rep} + U_{bond} + U_{prom}. \tag{8.12}$$

The repulsive energy $U_{rep}$ is assumed to be pairwise in character, as in eqn (4.29) or eqn (7.35). The covalent bond energy, following eqn (4.30) or eqn (7.32), is defined by

$$U_{bond} = \sum_{\alpha = s, p} \int^{E_F} (E - E_\alpha) n_\alpha(E) \, dE, \tag{8.13}$$

where $n_{s, p}(E)$ are the local s, p density of states, $E_{s, p}$ are the effective s, p atomic energy levels, and $E_F$ is the Fermi energy. In this section we will be considering only those lattices in which all sites are equivalent so that $n_\alpha(E)$ does not require a site-specific label. The densities of states will be evaluated assuming Harrison's (1980) parameterization for the bulk band structure, namely

$$\left.\begin{array}{r} pp\sigma(R) \\ pp\pi(R) \\ sp\sigma(R) \\ ss\sigma(R) \end{array}\right\} = \left.\begin{array}{r} -2.31 \\ 0.76 \\ -1.31 \\ 1.00 \end{array}\right\} h(R), \tag{8.14}$$

except that the value of $pp\pi$ in eqn (8.14) has been chosen 30% larger than Harrison's suggested value, in order to stabilize the close-packed structures with respect to the dimer for the case of the alkali metals with $N = 1$ (Cressoni and Pettifor (1991)). The promotion energy, $U_{prom}$, is given by

$$U_{prom} = (E_p - E_s)\Delta N_p = \Delta E_{sp}\Delta N_p, \tag{8.15}$$

where $\Delta N_p$ is the change in the number of p electrons between the free-atom ground state and the bulk.

The energy difference between two structure types is then given by the structural energy difference theorem as

$$\Delta U = [\Delta U_{bond} + \Delta U_{prom}]_{\Delta U_{rep} = 0}. \tag{8.16}$$

Defining the *band* energy as

$$U_{\text{band}} = \int^{E_F} E n(E) \, dE,$$   (8.17)

where $n(E)$ is the total s and p density of states per atom, the energy difference may be written as the difference in the band energy, namely

$$\Delta U = [\Delta U_{\text{band}}]_{\Delta U_{\text{rep}} = 0}.$$   (8.18)

Further, assuming that the repulsive pair potential varies as the square of the bond integrals, as in eqn (8.4), we have

$$\Delta U = [\Delta U_{\text{band}}]_{\Delta \mu_2 = 0}.$$   (8.19)

Thus, the energy difference between two structure types is given by the difference in their band energies, provided the bond integrals have been adjusted, so that each band has the same second moment or mean-square width. Taking the simple cubic lattice with $z = 6$ as reference with an equilibrium nearest-neighbour ss$\sigma$ bond integral $h_0$, then the appropriate bond integral $h_z$ for any other lattice with coordination $z$ will satisfy

$$z h_z^2 = 6 h_0^2.$$   (8.20)

Figure 8.4 shows the resultant pure s and hybridized sp band densities of states corresponding to zero sp splitting $\Delta E_{\text{sp}} = E_p - E_s = 0$ for structures with nearest-neighbour coordinations, $z = 2$ (the zig-zag linear chain with 90° bond angles), $z = 3$ (the single graphitic sheet or honeycomb lattice), $z = 4$ (both the diamond cubic and diamond hexagonal lattices), $z = 6$ (simple cubic), $z = 8$ (simple hexagonal), and $z = 12$ (both fcc and ideal hcp lattices). In addition, the densities of states are shown for the bcc lattice with $z = 14$ corresponding to eight first and six second nearest neighbours, where the bond integrals of the second nearest neighbour have been assumed to be one-third of the first nearest neighbours. The energy is measured in units of $|h_0|$, so that the simple cubic s band, for example, runs from $-6$ to $+6$ as expected. The sp densities of states for the linear chain ($z = 2$), the graphitic sheet ($z = 3$), and the diamond cubic lattice ($z = 4$) may be compared with our earlier simpler model with $\Delta E_{\text{sp}} = 0$ in the upper panel of Fig. 8.1. The influence of the bonding between the non-directed hybrids on neighbouring atoms is immediately apparent in the broadening of the earlier $\sigma$ bond energy levels, as too is the broadening of the non-bonding $\pi$ states for the linear chain and graphitic sheet. We see that a hybridization gap indeed opens up in the sp-valent diamond cubic and hexagonal densities of states. It is interesting to note that the pure s bands for the cubic and hexagonal diamond or close-packed lattices are identical as they have identical moments, which has been proved by Burdett and Lee (1985). It is the angular character of the valence orbitals which

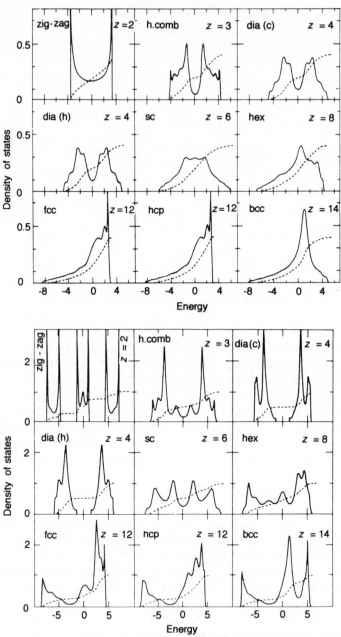

**Fig. 8.4** The s (upper panel) and sp (lower panel) density of states for different lattices in energy units of $|h_0|$ with $E_s = E_p = 0$. The broken curves give the integrated density of states, provided the numbers on the vertical scale are multiplied by five for the s case and eight for the sp case respectively. (From Cressoni and Pettifor (1991).)

distinguishes between cubic and hexagonal systems within a first nearest-neighbour TB model.

Figure 8.5 shows the structural energy for s-, p-, and sp-valent systems as a function of band filling which results from occupying the TB densities of states $n(E)$. The structural energy has been defined as the difference between the band energy for a given structure (cf eqn (8.17)) and that corresponding to a reference rectangular s, p, or sp density of states with the same second moment (cf the smooth parabolic variation of eqn (7.33)). This procedure allows the very small energy differences between the different structure types to be displayed more clearly. We see that, as the s band is filled with electrons, the most stable ground-state structure is predicted to change from close-packed → linear chain → dimer → linear chain → simple cubic. As the p band is filled with electrons, the most stable ground-state structure is predicted to change from fcc → hcp → fcc → puckered graphitic sheet → diamond cubic → liner chain → simple cubic → dimer → simple cubic, where the puckered graphitic sheet is a single three-fold coordinated layer of the arsenic structure with 90° bond angles (cf Fig. 1.5). As the sp band is filled with electrons, the most stable ground-state structure is predicted to change from fcc → hcp → fcc → simple hexagonal → graphitic sheet → diamond cubic → graphitic sheet → puckered graphitic sheet → linear chain → dimer.

The panels in Fig. 8.5 allow the rationalization of the structural trends that are observed within the sp-valent elements in Table 1.1. In particular, beginning on the right-hand side of the periodic table where the assumptions of the TB model are most appropriate, we see that the theory correctly predicts the dimeric structures of the halogens with $N = 7$ and the helical linear chain structures of the chalcogens with $N = 6$. The exceptions are oxygen with its dimeric behaviour and polonium with its simple cubic structure (sulphur exhibits structures based on helical chains at high temperatures). Nevertheless, we observe that both the dimeric and simple cubic structures are nearby in energy. The softer core of the 2p-valent oxygen atom would favour the lower coordinated structure type, whereas the larger sp splitting of the heavy polonium atom induced by relativistic effects and a harder core could favour the six-fold coordinated simple cubic lattice (which for the large sp splittings in the middle panel of Fig. 8.5 is stable for $6.3 < N < 6.8$).

The middle panel of Fig. 8.5 for large sp splittings shows that a single puckered graphitic sheet of the arsenic lattice is indeed the most stable structure for half-full p bands. This stability of the arsenic structure type for $N = 5$ persists almost to vanishingly small values of $\Delta E_{sp}$ as can be seen in the lowest panel of Fig. 8.5. The exceptions are nitrogen with its dimeric form, which is probably stabilized by its soft core, and phosphorus with its own variant of cutting three nearest-neighbour bonds within the simple cubic lattice (cf Fig. 1.5). The group IV elements are predicted to change from the open four-fold coordinated diamond cubic structure for small values of $\Delta E_{sp}$

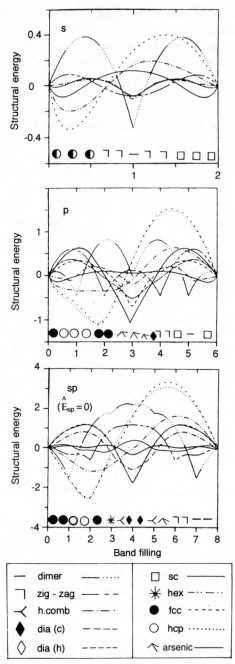

**Fig. 8.5** The structural energy in units of $|h_0|$ as a function of band filling for the pure s (upper panel), pure p (middle panel) and sp cases (lower panel). (After Cressoni and Pettifor (1991) and Cressoni (private communication).)

(lower panel) to the close-packed twelve-fold coordinated fcc structure for large values of $\Delta E_{sp}$ (middle panel). This is consistent with the observation in Table 1.1 that silicon, germanium, and tin are diamond cubic, whereas lead is fcc. The latter has a larger value of $\Delta E_{sp}$ than might be expected due to a 3 eV relativistic contribution which weakens the binding energy of the tetrahedral lattice due to an increased positive promotion energy (cf eqn (8.15)). Again the 2p element is exceptional with carbon taking the graphitic structure. Nevertheless, this latter structure is nearby in energy and could easily be stabilized by a softer core through a choice of $\alpha_h < \frac{1}{2}$. Finally, even though the sp-valent metals with $N = 1, 2$, and 3 are not accurately described by a first nearest-neighbour orthogonal TB model, we see that the simple model does predict correctly the occurrence of close-packed structures in this region.

## 8.4. Interpretation in terms of moments

The structural trends within the sp-valent elements are a direct result of the oscillatory behaviour of the energy curves displayed in Fig. 8.5. These oscillations can be understood in terms of the topology of the lattice by using the Ducastelle–Cyrot–Lackmann moments theorem that we introduced in Chapter 4. We saw that if two densities of states have moments that are identical up to some level, $p_0$ (i.e. $\Delta\mu_0 = \Delta\mu_1 = \cdots = \Delta\mu_{p_0} = 0$), then the energy difference as a function of band filling must have at least $(p_0 - 1)$ nodes over and above the zeroes corresponding to the empty or full band. Thus, since the reference rectangular density of states and all the crystal densities of states have the same values of $\mu_0$, $\mu_1$, and $\mu_2$ (the latter through the structural energy difference theorem, eqn (8.19)), any structural energy curve in Fig. 8.5 must cross zero at least once. A single crossing implies that the structural energy is dominated by a difference in the third moment from that of the uniform rectangular band, a double crossing implies a dominant difference in the fourth moment, a triple crossing indicates a dominant difference in the fifth moment, etc. Further, since the $p$th moment is related to closed paths of $p$ steps through eqn (4.47), one node will be related to closed paths of length three, two nodes to closed paths of length four, three nodes to closed paths of length five, etc.

This allows us to understand the oscillatory behaviour illustrated in Fig. 8.5. All those lattices which contain only even-membered rings will have all their odd moments identically zero (taking $E_s = E_p$ as the reference energy). Consequently, the densities of states and structural energy curves for the linear chain, graphitic layer, diamond, and simple cubic lattices will be *symmetric* as observed in Figs. 8.4 and 8.5. On the other hand, the simple hexagonal, fcc, hcp, and bcc structures have a sizeable third moment, $\mu_3$, due to the presence of odd three-membered rings within their lattices, so that their densities of states and structural energy curves will be *asymmetric* as

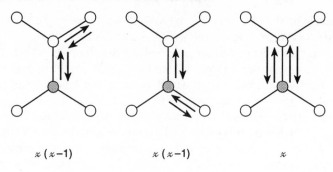

**Fig. 8.6** The number of different contributions to the fourth moment about a given atom on a Bethe lattice with local coordination, $z$.

observed in Figs. 8.4 and 8.5. Thus, the stability of close-packed structures over the more open structures for less than half-full bands is due to the presence of three-membered rings, which are absent in the latter structure types. We see in Fig. 8.5 that the difference in energy between the cubic and hexagonal close-packed structure types is very small, as expected. Moreover, since the curves cross three times for the sp-valent case, we deduce the importance of the fifth moment, $\mu_5$, in controlling the cubic versus hexagonal stability of close-packed sp-valent lattices.

The relative stability of structural types with even-membered rings can often be inferred from the dimensionless shape parameters, s, which from eqn (4.58) is given by

$$s = \hat{\mu}_4/\hat{\mu}_2^2 - 1, \tag{8.21}$$

since $\hat{\mu}_3 = 0$ (cf $\hat{\mu}_p = \mu_p/\mu_0$). We saw in Chapter 4 that if $s < 1$, the density of states shows bimodal behaviour, whereas if $s > 1$, the density of states shows unimodal behaviour. For the case of s orbitals, the fourth moment contribution for a Bethe lattice with no rings is easy to evaluate. From Fig. 8.6

$$\mu_4^{(B)} = [2z(z - 1) + z]h^4, \tag{8.22}$$

so that

$$s = 1 - z^{-1}. \tag{8.23}$$

This is the sole contribution for the dimer ($z = 1$), linear chain ($z = 2$), graphitic sheet ($z = 3$), and diamond lattice ($z = 4$), so that they take shape parameter values $s = 0, \frac{1}{2}, \frac{2}{3}$, and $\frac{3}{4}$ respectively. Their densities of states, therefore, will display bimodal behaviour as is observed in the upper panel of Fig. 8.4. The four-membered ring contributions play an important role in the close-packed lattices as can be seen from the upper curve in Fig. 8.7, which plots the values of $(s + 1)$, normalized by its value for the dimer,

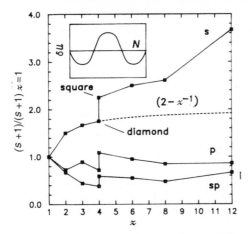

**Fig. 8.7** Plotting $(s+1)/(s+1)_{x=1}$ versus local coordination $x$ for the pure s and sp cases, where $s$ is the dimensionless shape parameter. The dashed curve gives the Bethe lattice result, eqn (8.23), in which there are no ring contributions. (After Cressoni and Pettifor (1991).)

versus the coordination, $x$. (Note that for s-valent orbitals $(s+1) = 1$ for $x = 1$). Therefore, the close-packed lattices, in particular, will display marked unimodal behaviour as is indeed apparent from the fcc, hcp, and bcc densities of states in Fig. 8.4. As expected, the dimer is the most stable structure for half-full bands because it displays the most bimodal behaviour $(s = 0)$, whereas the simple cubic lattice is the most stable structure for nearly-full bands because it displays the most unimodal behaviour $(s = 1.5)$ of the even-membered ring lattices.

The structural sequences in Fig. 8.5 change between the s-valent, p-valent and sp-valent cases because the shape parameter, s, is very sensitive to the angular character of the orbitals. We have already seen in Fig. 4.13 for the triatomic molecule $AH_2$ that the three-atom contributions to the fourth moment are dependent on the bond angle. It follows from eqs (7.14)–(7.17) that hopping along a three-atom path and back again amongst sp-valent orbitals will lead to the fourth-moment contribution

$$\mu_4^{(3)} = [ss\sigma^4 + sp\sigma^4 + pp\pi^4 + 2ss\sigma^2sp\sigma^2 + 2pp\pi^2(sp\sigma^2 + pp\sigma^2)]$$

$$+ 2sp\sigma^2(ss\sigma^2 - 2ss\sigma pp\sigma + pp\sigma^2) \cos\theta$$

$$+ [sp\sigma^4 + pp\sigma^4 + pp\pi^4 + 2sp\sigma^2pp\sigma^2 - 2pp\pi^2(sp\sigma^2 + pp\sigma^2)] \cos^2\theta.$$

$$(8.24)$$

Figure 8.8 shows the resultant angular dependence of this three-atom contribution for the s, p, and sp cases respectively. As expected, the s orbitals

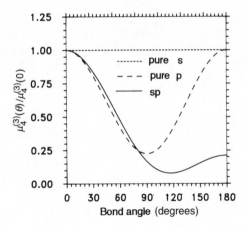

**Fig. 8.8** The angular dependence of the three-atom fourth moment contribution, eqn (8.24) for the pure s, pure p and sp cases. (From Cressoni and Pettifor (1991).)

display no bond-angle dependence because the only contribution from eqn (8.24) is the first term $ss\sigma^4$. On the other hand, the pure p case shows a marked minimum at 90°, since then

$$(\mu_4^{(3)})_p = pp\pi^2(pp\pi^2 + 2pp\sigma^2) + (pp\sigma^2 - pp\pi^2)^2 \cos^2\theta. \qquad (8.25)$$

Thus, we expect the puckered graphitic sheet with 90° bond angles to have the smallest normalized fourth moment and shape parameter, s, and hence to be the most stable structure for the half-full p band as is indeed observed in the middle panel of Fig. 8.5. We should also note that if the $\pi$ bonding is neglected then this three-atom contribution is identically zero for $\theta = 90°$, so that $s = 0$ and we have the total bimodal behaviour of the p eigenspectrum that is observed in the lower panel of Fig. 8.1 for the arsenic structure type.

For sp-valent orbitals, the minimum in Fig. 8.8 is skewed to higher angles, the minimum occurring for $\theta = 117°$ for the bond integrals chosen in eqn (8.14). This minimum is responsible for the diamond and graphite lattices with $\theta = 109°$ and 120° respectively having the smallest and second smallest values of the normalized fourth moment, and hence the shape parameter, s, in Fig. 8.7. This is reflected in the bimodal behaviour of their densities of states in Fig. 8.4 with a gap opening up for the case of the diamond cubic or hexagonal lattices. Hence, the diamond structure will be the most stable structure for half-full bands because it displays the most bimodal behaviour, whereas the dimer will be the most stable structure for nearly-full bands because it has the largest s value and hence the most unimodal behaviour of all the sp-valent lattices in Fig. 8.7. We expected to stabilize the graphitic structure as we move outwards from the half-full occupancy because this lattice has the second smallest value of $s$, whereas we expect to stabilize the linear chain as we move inwards from nearly-full occupancy because this lattice has the second largest value of $s$. This is indeed observed in Fig. 8.5. It is the same trend as displayed by the simpler model in the right-hand panel of Fig. 8.2 where it is easy to show that the corresponding values of $s$ for the diamond, graphitic, linear chain, and dimeric eigenspectra in the upper panel of Fig. 8.1 are $s = 0, \frac{1}{3}, 1$, and 3 respectively. Unlike the other lattices, the puckered graphitic layer and the simple cubic lattice have structural energy curves in Fig. 8.5 that cross each other four times, so that their relative stability is determined by the sixth moment. Their similarity in energy is not unexpected because they are two closely related structure types as indicated in Fig. 1.5.

## 8.5. Structural trends within the sd-valent elements

We saw in §7.5 that the rectangular d-band model was unable to differentiate between different structure types because it had approximated the true density of states by a constant value that was determined by the second moment, $\mu_2$, through the rectangular bandwidth, $W$. In practice, the densities of states of the transition metals display structure that is characteristic of the crystal lattice. In particular, the bcc density of states in Fig. 7.6 splits into definite bonding and antibonding regions around $N = 6$, corresponding to the band filling of bcc Cr, Mo, and W. On the other hand, the fcc and hcp densities of states are broadly similar as might be expected from their both having twelve equidistant nearest neighbours (assuming an ideal hcp lattice with an axial ratio, $c/a = \sqrt{8/3}$). The hcp density of states, unlike the fcc, does have, however, marked local minima around $N = 4$ and $N = 8$ corresponding to the band fillings of hcp Ti, Zr, and Hf and hcp Ru and Os respectively. Since this structure in the hybrid NFE–TB density of states reflects that of the pure TB d band, it is not surprising that the observed crystal structure sequence from hcp → bcc → hcp → fcc across the non-magnetic transition series is driven by the d bond contribution alone. As shown in Fig. 8.9 the structural trend is correctly reproduced apart from the noble metal end of the series where sp-d hybridization is required to obtain the correct fcc crystal structure of Ni, Pd, and Pt.

The strong stability of the bcc lattice with respect to either fcc or hcp for half-full bands is indicative of the bimodal behaviour that is displayed by the bcc density of states in Fig. 7.6. This can be tracked down to the differences in the four-membered ring contribution to the fourth moment, namely

$$\mu_4^{(4)} = \text{Tr} \quad H_{12} \quad H_{23} \quad H_{34} \quad H_{41} \tag{8.26}$$

where the $H_{ij}$ are the $5 \times 5$ TB matrices that link the five d orbitals on atom i to those on the neighbouring atom j. For the particular case of equilateral

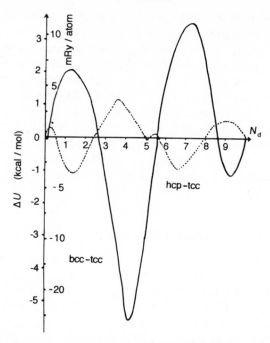

**Fig. 8.9** The d bond energy of the bcc (solid line) and the hcp (dotted line) lattices with respect to the fcc lattice as a function of the d band filling $N_d$. (From Pettifor (1972).)

planar rings with bond angle $\theta$

$$\mu_4^{(4)} = 5[(1757 - 60460x^2 + 327870x^4 - 563500x^6 + 300125x^8)/5792]h^4,$$

(8.27)

where $x = \cos\theta$ and $h^4 = (\text{dd}\sigma^4 + 2\text{dd}\pi^4 + 2\text{dd}\delta^4)/5$. This analytic expression was derived by Moriarty (1988) assuming the canonical parameters $\text{dd}\sigma:\text{dd}\pi:\text{dd}\delta = -6:4:-1$. The upper panel of Fig. 8.10 shows the resultant rapid oscillations as a function of bond angle. They reflect the interference between the angular lobes of the d orbitals as the electron hops around the ring from $1 \rightarrow 2 \rightarrow 3 \rightarrow 4 \rightarrow 1$. This angular prefactor would, of course, be totally absent for s orbitals, since then this four-membered ring contribution would be simply $\text{ss}\sigma^4$. The bcc and fcc planar rings, which are shown in the lower panel of Fig. 8.10, have bond angles of 70.5° (109.5°) and 90°, so that they provide negative and positive contribution to $\mu_4$ respectively. Consequently, we find $\mu_4^{\text{bcc}} < \mu_4^{\text{fcc}}$, which implies the greater bimodal behaviour of the bcc density of states. The difference in energy between the fcc and hcp lattices in Fig. 8.9 is about a factor of five smaller than the bcc–fcc energy difference and is more than two orders of magnitude smaller than the

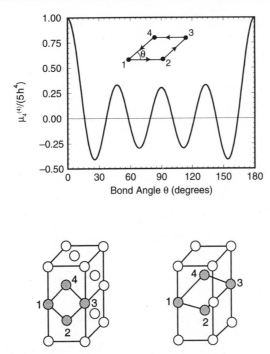

**Fig. 8.10** Upper panel: the four-membered ring contribution to the fourth moment of a d band as a function of the bond angle $\theta$. (After Moriarty (1988).) Lower panel: A four-membered ring contribution in the fcc and bcc lattices respectively. Note that from the upper panel the fcc and bcc rings shown contribute positive and negative contributions respectively to the fourth moment.

cohesive energy itself. The energy difference curve crosses zero essentially four times, so that cubic versus hexagonal close-packed stability is driven by the sixth moment $\mu_6$.

The trivalent rare-earth crystal structure sequence from hcp → Sm type → La type → fcc, which is observed for both decreasing atomic number and increasing pressure, is also determined by the d-band occupancy. Figure 8.11(a) shows the self-consistent LDA energy bands of fcc lanthanum as a function of the normalized atomic volume $\Omega/\Omega_0$, where $\Omega_0$ is the equilibrium atomic volume. We see that the bottom of the NFE sp band $\Gamma_1$ moves up rapidly in energy in the vicinity of the equilibrium atomic volume as the free electrons are compressed into the ion core region from where they are repelled by orthogonality constraints (cf eqn (7.29)). At the same time the d band widens, so that the number of d electrons increases under pressure

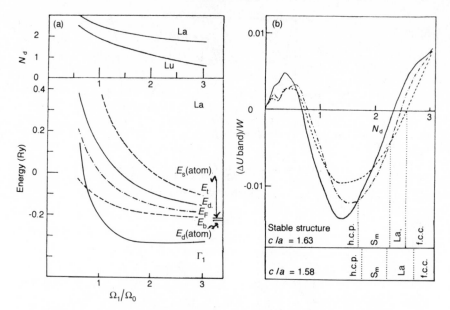

**Fig. 8.11** (a) The energy bands of La about the equilibrium atomic volume $\Omega_0$ and the corresponding d band occupancy, $N_d$, of La and Lu. $E_d$, $E_t$, and $E_b$ label the centre of gravity, the top and bottom of the d band respectively; $\Gamma_1$ is the bottom of the NFE sp band, and $E_F$ is the Fermi energy. (b) The relative d bond energies in units of the d bandwidth, $W$, of hcp (full curve), La structure type (dashed curve), and Sm structure type (dot-dashed curve) with respect to fcc as a function of the d band occupancy $N_d$. The resulting stable structures for the ideal and a non-ideal axial ratio are also shown. (From Duthie and Pettifor (1977).)

(as shown in the upper panel of Fig. 8.11(a)). Moreover, since La has a larger ion core than Lu (cf Table 2.1), the number of d electrons will also increase on moving from right to left across the rare earth series from Lu to La (as shown in the upper panel of Fig. 8.11(a)). This increase in the number of d electrons drives the structural trend from hcp → Sm type → La type → fcc as is demonstrated by Fig. 8.11(b), which compares the TB d bond energy of the four different close-packed lattices. We see, therefore, that the running of the string in Fig. 1.8 *backwards* through the periodic table from Lu to La is consistent with its direction through the transition elements, since both are in the direction of increasing $N_d$.

## 8.6. Anomalous structures due to magnetism

The presence of magnetism amongst the 3d transition elements causes magnanese, iron, and cobalt not to obey the structural trend that is observed across the nonmagnetic 4d and 5d series. Manganese takes the α-Mn

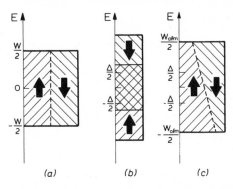

**Fig. 8.12** The rectangular d band model of the (a) nonmagnetic, (b) ferro-magnetic, and (c) antiferromagnetic states. (From Pettifor (1980).)

structure type with 58 atoms in the unit cell rather than the hcp lattice displayed by isovalent technetium and rhenium. Iron and cobalt take the bcc and hcp ground-state structures respectively, rather than the hcp and fcc lattices expected.

The magnetism amongst the 3d transition metals is well described by the theory of band magnetism, which was proposed by Stoner in 1939. As we have already seen in §3.4, a nonmagnetic system will become magnetic if the lowering in the exchange energy due to the alignment of the electron spins more than compensates for the corresponding increase in kinetic energy. This may be demonstrated by the rectangular d-band model of Fig. 8.12. In the nonmagnetic state, the up and down spin electrons are equivalent and, therefore, have identical density of states, $n_\uparrow$ and $n_\downarrow$, as shown in Fig. 8.12(a). In the magnetic state, the presence of a local magnetic moment, $m$, produces an exchange field $\Delta$ on the atom of strength

$$\Delta = Im, \tag{8.28}$$

where $I$ is the Stoner exchange integral and $m = N_d^\uparrow - N_d^\downarrow$ in Bohr magnetons ($\mu_B$) (cf eqn (3.40)). In the ferromagnetic state, all the atomic moments are aligned in the same direction, so that an up-spin electron sees the atomic level, $E_d$, shifted by $-\frac{1}{2}\Delta$ on every site and the down-spin electron shifted by $+\frac{1}{2}\Delta$. Therefore, the densities of states, $n_\uparrow$ and $n_\downarrow$, are shifted rigidly apart by $\Delta$, as shown in Fig. 8.12(b). On the other hand, in the antiferromagnetic state half the atoms have their moments aligned up and the other half have their moments aligned down, so that an electron sees two types of sites with energies $E_d \pm \frac{1}{2}\Delta$. The problem is, therefore, analogous to that of the AB alloy discussed in section 7.6 (cf Fig. 7.12) and the densities of states, $n_\uparrow$ and $n_\downarrow$ (corresponding to an atom with net moment up) are obtained by skewing the rectangular nonmagnetic densities of states as shown in Fig. 8.12(c).

The magnetic energy which accompanies the formation of a local moment,

$m$, at each site, may be written as

$$U_{mag} = \delta T - \tfrac{1}{4}Im^2 \qquad (8.29)$$

where the first term is the change in the kinetic energy and the second is the lowering in energy due to exchange. The ferromagnetic (fm) state is created by flipping $\tfrac{1}{2}m$ down-spin electrons from just below the nonmagnetic Fermi level into the unoccupied up-spin states just above the nonmagnetic Fermi level. This is accompanied by an increase in kinetic energy of $(\tfrac{1}{2}m)/n(E_F)$ per electron, so that to second order

$$U_{fm} = \tfrac{1}{4}m^2/n(E_F) - \tfrac{1}{4}Im^2 \qquad (8.30)$$

where in this present discussion $n(E_F)$ refers to the nonmagnetic density of states *per spin*. Therefore, the nonmagnetic state will be unstable to ferromagnetism if $U_{fm} < 0$, that is if

$$In(E_F) > 1, \qquad (8.31)$$

which is the famous *Stoner criterion*. The equilibrium value of $m$ in the ferromagnetic state is determined by the condition

$$I\overline{n(N_d, m)} = 1, \qquad (8.32)$$

where $\overline{n(N_d, m)}$ is the average of the nonmagnetic density of states per spin between the two energies corresponding to a band-filling of $n_\uparrow$ and $n_\downarrow$ respectively (see, for example, Gunnarsson (1976)).

The magnetic energy of the antiferromagnetic (afm) state can be obtained by adding up the band energies in Fig. 8.12(c) and subtracting off the exchange energy that has been double-counted, that is

$$U_{afm} = -\tfrac{1}{20}(W_{afm} - W)N_d(10 - N_d) - (-\tfrac{1}{4}Im^2), \qquad (8.33)$$

where from eqn (7.56)

$$W_{afm} = \{1 + 3(\Delta/W)^2\}^{1/2}W. \qquad (8.34)$$

Expanding eqn (8.34) to second order and using eqn (8.28), the nonmagnetic state is found to be unstable to antiferromagnetism if

$$I/W > [\tfrac{3}{10}N_d(10 - N_d)]^{-1}. \qquad (8.35)$$

This is the rectangular d-band model criterion equivalent to the exact second-order result, namely

$$I\chi_q(E_F) > 1, \qquad (8.36)$$

where $\chi_q(E_F)$ is the response function corresponding to the afm ordering wave vector, **q**. The usefulness of the present model is that eqs (8.33) and (8.34) include terms beyond the second order, so that the equilibrium value of the magnetic moment and energy may be obtained explicitly. Equation

(8.33) is stationary for

$$m = (1/\sqrt{3})\{[\tfrac{3}{10}N_{\mathrm{d}}(10 - N_{\mathrm{d}})]^2 - (W/I)^2\}^{1/2} \qquad (8.37)$$

when

$$U_{\mathrm{afm}} = [\tfrac{1}{20}WN_{\mathrm{d}}(10 - N_{\mathrm{d}})] - \tfrac{1}{6}W^2/I] - \tfrac{1}{4}Im^2. \qquad (8.38)$$

The first term in eqn (8.38) represents the change in kinetic energy, $\delta T$. The value of the moment given by eqn (8.37) is identical to that obtained by filling the up and down spin bands in Fig. 8.12(c) and solving eqn (8.28) self-consistently.

Figure 8.13 shows the regions of stability of the ferromagnetic and anti-ferromagnetic phases as a function of the normalized exchange integral, $I/W$, and band filling $N_{\mathrm{d}}$ for the rectangular d-band model. The ferromagnetic (fm) and antiferromagnetic (afm) phases are stable for values of $I/W$ above the critical curves, ABC (fm) and DBE (afm), which are defined by eqn (8.31) with $n(E_{\mathrm{F}}) = 5/W$ and eqn (8.35) respectively. In the region where both phases are stable, the ferromagnetic and antiferromagnetic states have the lower energy in region FBE and ABF respectively. The magnetic behaviour across the 3d series can be accounted for qualitatively by assigning the 3d transition metals values of $N_{\mathrm{d}}$ in Fig. 8.13, which fix nickel with 0.6 holes as found experimentally. The corresponding values of $I/W$ are marked by the crosses in Fig. 8.13, with their magnitude increasing across the series, since the bandwidth decreases from Cr to Ni just as observed in Table 7.1 for the corresponding 4d series from Mo to Pd. The Stoner exchange integral, $I$, is approximately constant across the series, taking the value, $I \simeq 1$ eV, so that $I/W = 0.2$ corresponds to a rectangular d bandwidth of 5 eV which is a reasonable estimate for iron (cf Fig. 7.5).

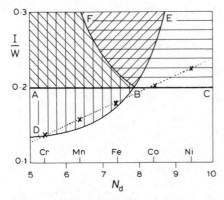

**Fig. 8.13** The regions of stability of the ferromagnetic and antiferromagnetic states as a function of the normalized exchange integral $I/W$ and d band filling $N_{\mathrm{d}}$. The crosses mark plausible values of $I/W$ across the 3d series. (From Pettifor (1980).)

The positions of the crosses in Fig. 8.13 imply that Cr, Mn, and Fe are antiferromagnets with moments of 0.7, 1.6, and $0.9\mu_B$ respectively, whereas cobalt and nickel are ferromagnets with moments of 1.6 and $0.6\mu_B$ respectively (resulting from fully occupying the up-spin d band with five electrons). This is in reasonable agreement with experiment where bcc Cr and fcc Fe are antiferromagnets with moments of about $0.6\mu_B$, whereas hcp cobalt and fcc nickel are ferromagnets with moments of 1.7 and $0.6\mu_B$ respectively. Although the Stoner criterion for ferromagnetism is not satisfied within the rectangular d band model for iron (cf Fig. 8.13), the nonmagnetic density of states of bcc iron has a large peak at the Fermi energy (cf Fig. 7.6), thereby driving the bcc lattice ferromagnetic.

The anomalous crystal structures of iron and cobalt may now be understood. Assuming that the up-spin d band in the ferromagnetic state is full with 5 electrons (cf Fig. 8.12(b)), then the down-spin d band will contain 2.4 and 3.4 electrons in iron and cobalt respectively, taking the 7.4 and 8.4 d electron total assigned in Fig. 8.13. Therefore, the up-spin electrons contribute nothing to the d-bond energy. The down-spin electrons, on the other hand, are equivalent to a fractional d-band occupancy of 4.8/10 and 6.8/10 respectively, so that from Fig. 8.9 they will drive the ferromagnetic iron lattice bcc and ferromagnetic cobalt lattice hcp as is observed experimentally.

In fact, iron exhibits all three common metallic crystal structures bcc, fcc, and hcp within its pressure–temperature phase diagram, as is shown by the inset of Fig. 8.14. The transition from the bcc $\alpha$ phase to the hcp $\varepsilon$ phase

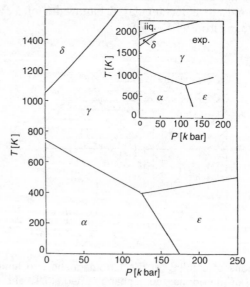

**Fig. 8.14** The theoretical phase diagram of iron compared with the experimental (inset). (From Hasegawa and Pettifor (1983).)

**Table 8.1** Coordination polyhedra of the $\alpha$-Mn structure type

| Site | Number of atoms per unit cell | Local coord. polyhedron | Magnetic moment | Size of atom |
|------|------|------|------|------|
| I | 2 | 16 | Large | Large |
| II | 8 | 16 | Large | Large |
| III | 24 | 13 | Small | Small |
| IV | 24 | 12″ | Small | Small |

under pressure follows from eqn (8.30) because the stabilizing magnetic energy of the bcc lattice with respect to the nonmagnetic hcp lattice falls as the density of states at the Fermi level decreases with increasing bandwidth. The transition from the bcc $\alpha$ phase to the fcc $\gamma$ phase as the temperature is increased is due to large magnetic fluctuations building up within the iron atoms on the fcc lattice, which can be modelled qualitatively by a finite temperature theory of band magnetism (see Fig. 8.14). The return to the bcc $\delta$ phase at still higher temperatures is driven by the additional magnetic entropy for the bcc lattice that arises from the gradual destruction of the short-range magnetic order as the temperature increases above the Curie temperature (Kaufman *et al.* (1963)).

The anomalous ground-state structure of elemental manganese is also stabilized by magnetism. The $\alpha$-Mn structure is taken by alloys of the isovalent 4d and 5d elements, technetium and rhenium, with transition elements to their left in the periodic table, the so-called $\chi$ phases, such as $Nb_{0.25}Tc_{0.75}$, $Zr_{0.14}Tc_{0.86}$, $Ta_{0.25}Re_{0.75}$ and $Hf_{0.14}Re_{0.86}$. The structure is characterized by four non-equivalent sites as listed in Table 8.1. Amongst the binary $\chi$-phases the two sites with 16-fold coordination are found to contain on average the larger alloying atoms to the left of Tc and Re, whereas the 12-fold and 13-fold coordinated sites contain on average the smaller majority atoms Tc or Re. Elemental manganese, on the other hand, has atoms with large moments situated at the centre of the 16-fold coordination polyhedra and atoms with small moments at the centre of the 12- and 13-fold coordination polyhedra. This is consistent with the fact that the onset of magnetism is accompanied by an expansion of the lattice that scales roughly as the square of the local magnetic moment. Thus, the presence of magnetism allows elemental manganese to behave like the neighbouring binary $\chi$-phases, where the atomic size difference between the constituent atoms helps stabilize this structure type for favourable electron-per-atom ratios or band filling.

## 8.7. Structural trends within the pd-bonded AB compounds

The atomic size difference plays an important role in helping to stabilize different structure types within binary systems. This is most easily demonstrated by considering the relative stability of the NaCl, CsCl, and cubic ZnS structure types within a hard-core ionic model, in which all the bonding arises from the electrostatic interaction between the positive and negative ion cores. These long-ranged coulomb interactions may be summed over all the lattice to give the Madelung energy per AB unit, namely

$$U_{\text{Madelung}} = -\alpha(Z^2e^2/4\pi\varepsilon_0 R),  \tag{8.39}$$

where $R$ is the nearest-neighbour distance, $\alpha$ is the Madelung constant, taking the values $\alpha = 1.748$, $1.763$, and $1.638$ for the NaCl, CsCl, and cubic ZnS lattices respectively (see, for example, Kittel (1986)).

The relative stability of the different structure types may now be calculated as a function of the radius ratio, $R_+/R_-$, where $R_+$ and $R_-$ are the hard-core radii of the positive and negative ions respectively. Figure 8.15 shows that when the ions are of equal size, the CsCl lattice with eight nearest neighbours and the largest Madelung constant is the most stable. However, as the radius ratio decreases, the structural trend from CsCl → NaCl → ZnS is found. The structural transition from CsCl to NaCl is a direct consequence of the fact that the volume of the CsCl lattice is determined solely by the second nearest-neighbour anion–anion interactions for

$$R_+/R_- < 3^{1/2} - 1 = 0.732,  \tag{8.40}$$

rather than by the first nearest-neighbour anion–cation interaction. For small values of $R_+$ satisfying eqn (8.40) the cations rattle around inside

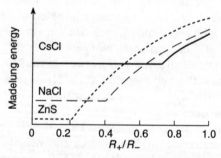

**Fig. 8.15** The Madelung energy in ionic compounds as a function of the radius ratio for CsCl, NaCl and cubic ZnS lattices (assuming the anion radius, $R_-$, is held constant).

the simple cubic cage of anions, so that the volume and hence the Madelung energy is constant (for fixed $R_-$). This accounts for the discontinuity at 0.732 in the CsCl curve in Fig. 8.15. A similar discontinuity in the NaCl curve occurs for the much smaller ratio of

$$R_+/R_- = 2^{1/2} - 1 = 0.414. \tag{8.41}$$

In practice, this hard-core model is too simple to predict reliably the ground-state structure of ionic compounds such as the alkali halides that are located in the upper left-hand corner of the AB structure map in Fig. 1.9. Nevertheless, it provides a simple introduction to the importance of the radius ratio in determining structural stability.

Most of the binary compounds within the AB structure map are not good insulators that are held together by ionic bonds but are metals or semi-conductors whose bonding is well described by the TB model. In this section we will, therefore, consider the relative stability of the seven most frequently occurring structure types amongst the pd-bonded AB compounds that are displayed in the upper panel of Fig. 8.16. The explicit contribution of the valence-s electrons will be neglected, so that we need only consider the bonding between the valence p and d states on the metalloid and transition element sites respectively. Within the canonical TB theory of Andersen (1975) the bond integrals are given explicitly by

$$dd(\sigma, \pi, \delta) = (-6, 4, -1)(r_d/R)^5 \tag{8.42}$$

$$pp(\sigma, \pi) = (2, -1)(r_p/R)^3 \tag{8.43}$$

$$pd(\sigma, \pi) = (-3, 3^{1/2})(r_p^3 r_d^5)^{1/2}/R^4, \tag{8.44}$$

where $R$ is the internuclear separation and $r_p$ and $r_d$ are constants character-istic of the particular A and B constituents. Choosing a degree of normalized hardness of the potential, $\alpha_h = \frac{1}{2}$, the repulsive pair potentials fall off with distance as the square of the corresponding bond integrals, that is

$$\Phi_{dd}(R) = C_d^2/R^{10} \tag{8.45}$$

$$\Phi_{pp}(R) = C_p^2/R^6 \tag{8.46}$$

$$\Phi_{pd}(R) = [\Phi_{pp}(R)\Phi_{dd}(R)]^{1/2} = C_p C_d/R^8 \tag{8.47}$$

where $C_p$ and $C_d$ are constants characteristic of the particular A and B constituents.

The structural energy difference theorem requires that we prepare the lattices so that they all display the same repulsive energy. It follows from eqs (8.45)–(8.47) that the repulsive energy per AB unit may be written

$$U_{rep} = (C_p C_d/\Omega_{AB}^{8/3})(\alpha_{pd} + \tfrac{1}{2}\mathscr{R}^{-1}\alpha_{dd} + \tfrac{1}{2}\mathscr{R}\alpha_{pp}) \tag{8.48}$$

where

$$\alpha_{ll'} = \mathscr{N}^{-1} \sum (\Omega_{AB}^{1/3}/R)^{2(l+l'+1)} \tag{8.49}$$

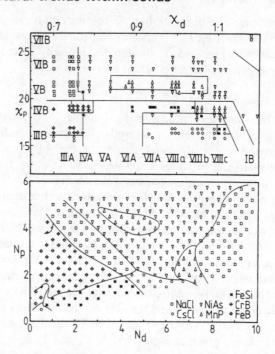

**Fig. 8.16** (The upper panel shows the structure map ($\chi_p$, $\chi_d$) for 169 pd bonded AB compounds, where $\chi_p$ and $\chi_d$ are values for the A and B constituents of a certain chemical scale $\chi$ which orders the elements in a similar way to the relative ordering number, $\mathcal{M}$. The lower panel shows the theoretical structure map ($N_p$, $N_d$), where $N_p$ and $N_d$ are the number of p and d electrons respectively. (From Pettifor and Podloucky (1984).)

with $\alpha_{dd} = \alpha_{22}$, $\alpha_{pp} = \alpha_{11}$, and $\alpha_{pd} = \alpha_{12}$ and with the sum in eqn (8.49) extending over the relevant dd, pp, or pd interactions on the lattice. The quantity, $\Omega_{AB}$, is the volume per AB unit. The *relative size factor*, $\mathcal{R}$ is defined by

$$\mathcal{R} = (C_p/C_d)\Omega_{AB}^{2/3}, \tag{8.50}$$

where the volume dependence enters as the result of the different distance behaviour of the p and d potentials $\Phi_{pp}$ and $\Phi_{dd}$ respectively. It is a measure of the relative size of the p and d atoms as can be seen as follows. Consider a pair of transition-metal atoms a distance $2R_d = \Omega_{AB}^{1/3}$ apart. Then from eqs (8.45) and (8.46) a pair of metalloid atoms will show the same repulsive energy when they are separated by the distance $2R_p$ such that

$$(R_p/R_d)^3 = \mathcal{R}. \tag{8.51}$$

Thus, $\mathcal{R}$ generalizes the concept of the relative size of hard spheres to the case where the mutual interaction varies smoothly rather than discontinuously.

The repulsive coefficients, $\alpha_{ll'}$, depend only on structure and not on volume. They are given in Table 8.2, where the measured internal coordinates of MnP, FeB, CrB, and FeSi have been used to characterize the corresponding structure types. Two choices of the axial ratio, $c/a$, have been chosen for the NiAs structure types; the first corresponds to the observed ratio of 1.39 for the NiAs compound, the second to the ideal ratio of $(8/3)^{1/2}$. The repulsive potential in eqn (8.49) has been summed over the nearest neighbours only, other than for the pp coefficients of CsCl and FeSi, where further neighbours were included to generate better than 5% accuracy. The values of the repulsive coefficients in Table 8.2 reflect the local atomic environment. The metal (M) and metalloid (X) sites on the cF8 (NaCl) 6/6 lattice are both sixfold octahedrally coordinated with X and M atoms respectively, which form two interpenetrating cubic close-packed lattices. NaCl can, therefore, be described as a *cubic* close-packed arrangement of X atoms with all the octahedral holes occupied by M atoms. Structure type hP4 (NiAs) $8^{IV}/6'$, on the other hand, consists of a *hexagonal* close-packed arrangement of X atoms with the octahedral holes occupied by M atoms as shown in Fig. 1.10. The X sites are, thus, sixfold coordinated by a trigonal prism of M atoms, which themselves reside on a simple hexagonal lattice with eight nearest neighbours. The oP8 (MnP)$10'''/8'''$ structure type is an orthorhombically distorted

**Table 8.2** The repulsive coefficients $\alpha_{ll'}$. The number in brackets gives the number of atoms included in the repulsive sum in eqn (8.49). (From Pettifor and Podloucky (1986).)

|  | $\alpha_{pd}$ | $\alpha_{dd}$ | $\alpha_{pp}$ |
|---|---|---|---|
| NaCl | 38.2 | 3.8 | 6.0 |
|  | (6) | (12) | (12) |
| NiAs | 36.5 | 15.2 | 6.2 |
| ($c/a = 1.39$) | (6) | (8) | (12) |
| NiAs | 38.2 | 6.7 | 6.0 |
| ($c/a = 1.633$) | (6) | (8) | (12) |
| MnP | 33.3 | 7.8 | 10.0 |
|  | (6) | (8) | (12) |
| CsCl | 25.2 | 6.0 | 7.5 |
|  | (8) | (6) | (6 + 12) |
| FeSi | 26.6 | 7.9 | 7.7 |
|  | (7) | (6) | (6 + 12) |
| FeB | 25.2 | 5.1 | 19.4 |
|  | (7) | (10) | (10) |
| CrB | 24.0 | 5.2 | 22.4 |
|  | (7) | (10) | (10) |

variant of NiAs in which the X and M sites are surrounded by distorted trigonal prisms and octahedra respectively (cf. Fig. 1.9). The metal and metalloid sites on the cP2 (CsCl) 14/14 structure type are both surrounded by eight unlike atoms at the corners of a cube. The like atoms lie on a simple cubic lattice with six first- and twelve second-nearest neighbours respectively. The cP8 (FeSi) 13'/13' structure type is related to CsCl but has seven nearest neighbours of opposite type instead of eight (cf. Fig. 1.9). The very similar oP8 (FeB) 17/9 and oC8 (CrB) 17/9 structure types are built up from zigzag chains of X atoms, which are surrounded by trigonal prisms of M atoms as shown in Fig. 1.12. The stacking together of these chains leads to an additional metal atom entering the coordination shell of the X site so that there are seven unlike nearest neighbours.

The relative prepared volumes, which are required to satisfy the structural energy difference theorem, may be obtained from eqn (8.48) by setting $\Delta U_{rep} = 0$. The resultant first-order change in volume $\Delta\Omega$ is given by

$$\frac{\Delta\Omega}{\Omega} = \frac{6\Delta\alpha_{pd} + 3\mathscr{R}^{-1}\Delta\alpha_{dd} + 3\mathscr{R}\Delta\alpha_{pp}}{16\alpha_{pd} + 10\mathscr{R}^{-1}\alpha_{dd} + 6\mathscr{R}\alpha_{pp}}, \tag{8.52}$$

where the $\Delta\alpha_{ll'}$ are the corresponding changes in the repulsive coefficients. Therefore, the fractional change in volume is a function only of the relative size factor, $\mathscr{R}$, so that a universal curve may be plotted for each structure with a given set of internal coordinates. Figure 8.17 shows these curves with the CsCl lattice as reference, using the values of the repulsive coefficients in Table 8.2. As expected, the NaCl lattice has the smallest volume at either end of the $\mathscr{R}$ scale because, as $C_p$ or $C_d$ tends to zero, the repulsion is dominated by one or other of the close-packed fcc sublattices. On the other hand, in the middle of the scale where the nearest-neighbour pd repulsion dominates, the volume of the NaCl lattice with six nearest neighbours is about 13% larger than the CsCl with eight nearest neighbours. The packing of hard spheres would have predicted the much larger volume difference of 30%.

The upper NiAs curve in Fig. 8.17, which corresponds to the observed axial ratio of 1.39 for NiAs, shows an increase in fractional volume with decreasing $\mathscr{R}$ due to the large value of $\alpha_{dd}$ in Table 8.2. This arises from the strong dd repulsion within the chain of atoms lying along the $c$-axis of the simple hexagonal metal sublattice. If the small Ni atoms are replaced by the larger transition metal atoms such as Ti or V from near the beginning of the series, then the axial ratio is observed to increase markedly in order to accommodate the increased dd repulsion. The lower NiAs curve is drawn for $c/a = 1.633$, with this ideal axial ratio being adopted by MnTe. We see from Table 8.2 that $\alpha_{dd}$ now assumes a value more typical of the other structure types. The FeB, CrB curve in Fig. 8.17 is very strongly dependent on $\mathscr{R}$ as a result of the strong pp repulsion along the zigzag chains of p

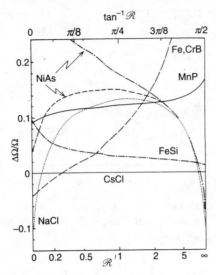

**Fig. 8.17** The fractional change in volume $\Delta\Omega/\Omega$ with respect to the CsCl lattice versus the relative size factor, $\mathcal{R}$. The upper and lower NiAs curves correspond to $c/a = 1.39$ and $(8/3)^{1/2}$ respectively. (From Pettifor and Podloucky (1984).)

atoms that thread the boride lattices. This is reflected in the anomalously large value of $\alpha_{pp}$ for FeB and CrB in Table 8.2.

The structural stability of the pd-bonded AB compounds may now be predicted by comparing the TB bond energy of the different lattices at the volume determined by the relative size factor, $\mathcal{R}$. Apart from the borides and NiAs, the fractional volume changes in Fig. 8.17 are not too sensitive to the particular choice of $\mathcal{R}$ for $0.3 < \mathcal{R} < 3$, so that the value $\mathcal{R} = 0.8$ has been chosen that is characteristic of the 4d and 5p elements. Figure 8.18 shows the resultant structural energies as a function of band filling $N$ for the case where the atomic p level on the A site and the atomic d level on the B site are equal, that is $\Delta E_{pd} = E_d - E_p = 0$. As $N$ increases we find the structural sequence from CsCl $\to$ FeSi $\to$ CrB $\to$ NaCl $\to$ NiAs $\to$ NaCl. The low band-filling stability of the CsCl, FeSi, and CrB lattices results from a sizeable third moment, $\mu_3$, which arises from the presence of many three-membered rings in these close-packed structure types. On the other hand, the absence of nearest-neighbour three-membered rings in the more open NaCl and NiAs structure types accounts for their stability for $N \geq 5.5$. The minimum in the NaCl and NiAs curves at $N \approx 6$ corresponds to a minimum in their density of states at the Fermi energy when all the pd-bonded orbitals are occupied.

The structural energy curves depend not only on the electron-per-atom

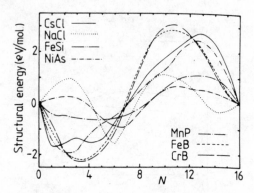

**Fig. 8.18** The structural energy as a function of band filling $N$ of the seven different structure types with $\Delta E_{pd} = 0$. (From Pettifor and Podloucky (1984).)

ratio or $N$ but also on $\Delta E_{pd} = E_d - E_p$, which is a measure of the Mulliken electronegativity difference. However, rather than plotting the most stable predicted structure on a structure map of $\Delta E_{pd}$ versus $N$, the rotated frame of $N_p$ versus $N_d$ is used instead in order to make direct comparison with the experimental results in the upper panel of Fig. 8.16. The terms, $N_p$ and $N_d$, stand for the number of p and d valence electrons associated with atoms A and B respectively. The resulting theoretical structure map is shown in the lower panel of Fig. 8.16. We observe that the structural sequence from CsCl → FeSi → CrB → NaCl → NiAs → NaCl is found along the diagonal from the lower left-hand corner to the upper right-hand corner, which is consistent with Fig. 8.18 for $\Delta E_{pd} = 0$.

We see that the TB model predicts the broad topological features of the experimental pd-bonded AB structure map. In particular, NaCl, situated in the top left-hand corner of the map adjoins NiAs running across to the right and boride stability running down to the bottom. The stability of MnP is found in the middle of the NiAs domain and towards the bottom right-hand corner, where it joins CsCl towards the bottom. The main failure of this simple pd TB model is its inability to predict the FeSi stability of the transition-metal silicides, which is probably due to the neglect of the valence s electrons within the bonding. Thus, the observed stability of the NaCl, CsCl, NiAs, MnP, and boride domains amongst the pd-bonded AB compounds is determined solely by the quantum mechanical covalent bond energy, once the bond lengths have been adjusted to account for atomic size differences. The classical electrostatic Madelung energy plays no role in these metallic systems because the atoms are perfectly screened and, hence, locally charge neutral.

## 8.8.  Bond order potentials

The link between the oscillatory behaviour of the structural energy curves
and the moments of the local density of states can be made explicit by writing
the bond order of a given bond as a many-atom expansion about that bond
(Pettifor (1989)). Considering for simplicity the case of s orbitals, on a lattice
where all sites are equivalent, the bond order can be expressed exactly (Aoki
(1993)) as

$$\Theta = \frac{2}{\sqrt{z}} \{\hat{\chi}_2(N) + \hat{\chi}_3(N)(\hat{\mu}_3/\hat{\mu}_2^{3/2}) + \hat{\chi}_4(N)[\hat{\mu}_4/\hat{\mu}_2^2 - \hat{\mu}_3^2/\hat{\mu}_2^3 - 2] + \cdots\},$$

$$(8.53)$$

where $z$ is the coordination of the lattice and $\hat{\chi}_n(N)$ represent normalized
response functions that depend on both the band filling, $N$, and the local
atomic environment through the moments, $\mu_p$. If the local density of states
falls on the unimodal–bimodal boundary with the dimensionless shape
parameter, $s = 1$, then the response functions take the particularly simple
form

$$\hat{\chi}_n(N) = \frac{2}{\pi} \left[ \frac{\sin(n-1)\phi_F}{n-1} - \frac{\sin(n+1)\phi_F}{n+1} \right], \qquad (8.54)$$

where $\phi_F$ is related to the band filling, $N$, through

$$N = (2\phi_F/\pi)[1 - (\sin 2\phi_F)/2\phi_F].$$

These response functions are plotted in Fig. 8.19 where we see that the
number of nodes (excluding the end points) equals $(n-2)$.

The many-atom expansion for the bond order, therefore, formulates
mathematically what we have already found earlier regarding the close link

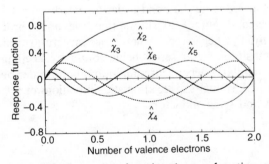

**Fig. 8.19** The normalized response function $\hat{\chi}_n$ as a function of the number of
valence s electrons per atom for the case of the dimensionless shape parameter,
$s = 1$. (From Pettifor and Aoki (1991).)

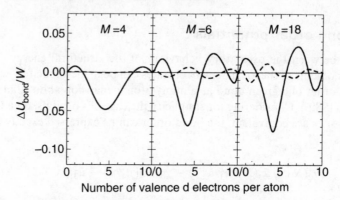

**Fig. 8.20** Convergence of the bcc–fcc d bond energy (full curve) and hcp–fcc d bond energy (dashed curve) with respect to the number of terms in the bond order potential expansion. The left, middle and right panels correspond to keeping terms up to fourth, sixth and eighteenth moments respectively. (From Aoki (1993).)

between the oscillatory behaviour of the structural energy curves and the moments of the local density of states. The first term in eqn (8.53) corresponds to our previous bond order potential expression, eqn (8.11), if $\hat{\chi}_2(N) = 1$. This term reflects the second moment of the density of states and gives rise to the inverse square-root dependence on the local coordination that is outside the curly brackets in eqn (8.53). The second term in the many-atom expansion results from the skewness of the bands, $\hat{\mu}_3/\hat{\mu}_2^{3/2}$, so that the prefactor, $\hat{\chi}_3(N)$, increases the bond order for less than half-full bands but reduces it for more than half-full (cf Fig. 8.19). The third term in eqn (8.53) reflects the unimodal–bimodal behaviour of the bands, since the moment expression inside the square brackets is $(s - 1)$, where $s$ is the dimensionless shape parameter, eqn (4.58). Thus, for $s < 1$ the prefactor $\hat{\chi}_4(N)$ will lead to this contribution enhancing the bond order for nearly half-full bands but reducing the bond order for nearly empty or nearly full bands, as expected. Figure 8.20 shows that the bond order expansion converges rapidly, the hcp → bcc → hcp → fcc structural trend across the nonmagnetic transition metal series being found by keeping terms only up to $\mu_6$. These angularly dependent many-atom bond order potentials may thus be used for performing realistic atomistic simulations of defect behaviour in intermetallics and semiconductors.

## References

Andersen, O. K. (1975). *Physical Review* **B12**, 3060.
Aoki, M. (1993). *Physical Review Letters* **71**, 3842.
Burdett, J. K. and Lee, S. (1985). *Journal of the American Chemical Society* **107**, 3063.

Cressoni, J. C. and Pettifor, D. G. (1991). *Journal of Physics: Condensed Matter* **3**, 495.

Ducastelle, F. and Cyrot-Lackmann, F. (1971). *Journal of Physics and Chemistry of Solids* **32**, 285.

Duthie, J. C. and Pettifor, D. G. (1977). *Physical Review Letters* **38**, 564.

Gunnarsson, O. (1976). *Journal of Physics* **F6**, 587.

Harrison, W. A. (1980). *Electronic Structure and the Properties of Solids*. Freeman, San Francisco.

Hasegawa, H. and Pettifor, D. G. (1983). *Physical Review Letters* **50**, 130.

Kaufman, L., Clougherty, E. V. and Weiss, R. J. (1963). *Acta Metallurgica* **11**, 323.

Kittel, C. (1986). *Introduction to Solid State Physics*. Wiley, New York.

Moriarty, J. A. (1988). *Physical Review* **B38**, 3199.

Nishitani, S. R., Alinaghian, P., Hausleitner, C., and Pettifor, D. G. (1994). *Philosophical Magazine Letters* **69**, 177.

Pettifor, D. G. (1972). *Metallurgical Chemistry* (ed. O. Kubaschewski), p. 191. Her Majesty's Stationery Office, London.

Pettifor, D. G. (1980). *Journal of Magnetism and Magnetic Materials* **15–18**, 847.

Pettifor, D. G. (1989). *Physical Review Letters* **63**, 2480.

Pettifor, D. G. and Aoki, M. (1991). *Philosophical Transactions of the Royal Society London* **A334**, 439.

Pettifor, D. G. and Podloucky, R. (1984). *Physical Review Letters* **53**, 1080.

Pettifor, D. G. and Podloucky, R. (1986). *Journal of Physics* **C19**, 315.

# Problems

*The first digit in the question number gives the corresponding chapter to which the problem relates.*

1.1 Plot the nearest neighbour histogram, showing the number of atoms in a given neighbouring shell versus shell distance, up to and including fourth nearest neighbours for the following structure types: (i) face-centred cubic; (ii) body-centred cubic; (iii) hexagonal close-packed with ideal axial ratio; and (iv) diamond. Hence, using the maximum-gap rule assign the appropriate nearest neighbour atoms to the local coordination polyhedron of each structure type, sketch and label with the relevant Jensen notation. Show that the fcc and ideal hcp histograms differ only beyond the second shell of neighbours.

Discuss what is meant by a Bravais lattice and the Pearson notation. Sketch the unit cells for fcc, bcc, hcp and diamond giving the corresponding Pearson notation.

1.2 (i) NaCl and NiAs are the first and fifth most frequently occurring AB structure types respectively. By referring to Fig. 1.10 explain why the key to the AB structure map labels them cF8 (NaCl) 6/6 and hP4 (NiAs) $8^{IV}/6'$ respectively. Comment on why it is not unexpected that the small TiAs and NbAs domains are located at the boundary between the main NaCl and NiAs domains within the AB structure map.

(ii) CsCl, TiCu and NaTl are three different ordered structure types with respect to an underlying bcc lattice. By referring to Fig. 1.11 explain why they are assigned the Pearson notation cP2, tP4 and cF16 respectively. Locate the small domains of TiCu and NaTl stability in the AB structure map and show that they occur, as expected, near domains of CsCl stability. Show that the Tl sites in NaTl form the tetrahedrally coordinated diamond lattice. Hence, give an explanation for the occurrence of this Zintl phase by assuming that all the valence electrons transfer from the less electronegative to the more electronegative sites.

2.1 Describe how the photoelectric effect and electron diffraction demonstrate the particle-like character of radiation and the wave-like character of particles respectively. Show how Heisenberg's uncertainty principle embraces the concept of wave–particle duality.

2.2   An electron, moving in one dimension, is confined by rigid walls to
the region $0 < x < L$. By solving the Schrödinger equation find the
eigenvalues $E_n$ and the eigenfunctions $\psi_n$. Show that the eigenfunctions
are orthogonal in that the overlap integral between the different states
vanishes, i.e.

$$\int_0^L \psi_n \psi_m \, dx = 0 \qquad \text{for } n \neq m.$$

Evaluate the first excitation energy of the electron from its ground state
for the cases $L = 1$ Å (an atomic sized 'quantum-dot') and $L = 1$ cm
(a short metallic pin), giving the answers in eV.

   Estimate the ground state energy using Heisenberg's uncertainty
principle and compare with the exact result.

2.3   The dominant contributions to the binding energy per electron of
jellium can be written in the form

$$U = A/r_s^2 - B/r_s$$

where $A$ and $B$ are constants and $r_s$ is the radius of the sphere containing
one electron on average. Show that the two terms correspond to the
repulsive kinetic energy of a free electron gas and the attractive potential
energy arising from the exchange-correlation hole respectively. Evaluate
the constants $A$ and $B$ (in atomic units) assuming for the latter case
that one electron is excluded from a sphere of radius $r_s$ about the given
electron. Find the value of the effective electronic radius and binding
energy at equilibrium and compare with the values of 4.0 au and 1.1 eV,
respectively, for metallic sodium.

2.4   The radial functions for the 1s, 2s, and 2p states of the hydrogen atom
are given by

$$R_{1s}(r) = 2e^{-r},$$

$$R_{2s}(r) = \frac{1}{\sqrt{2}}(1 - \tfrac{1}{2}r)e^{-r/2},$$

$$R_{2p}(r) = \frac{1}{\sqrt{24}} r\, e^{-r/2}.$$

Plot these radial functions and the corresponding radial probability
densities. By differentiation of the radial probability density show that
the maximum probability of locating the electron a distance $r$ from the
nucleus occurs at the first and second Bohr radii for the 1s and 2p
states respectively. Where does the maximum probability occur for the
2s state? Why does the 2s radial function have one node but the 2p
radial function is nodeless (outside the origin)?

Show that the average or expectation values of the radial distance $r$ for these three radial functions satisfy

$$\bar{r}_{nl} = [1 + \tfrac{1}{2}(1 - l(l+1)/n^2)]n^2$$

3.1 Show that the LCAO secular equation for the AB dimer can be written in the form of eqn (3.16), namely

$$\begin{pmatrix} -\tfrac{1}{2}\Delta E - (E - \bar{E}) & h - (E - \bar{E})S \\ h - (E - \bar{E})S & \tfrac{1}{2}\Delta E - (E - \bar{E}) \end{pmatrix}\begin{pmatrix} c_A \\ c_B \end{pmatrix} = 0$$

stating clearly any approximations made and the significance of the terms $h$, $S$, and $\Delta E$. Hence, show that the bonding eigenfunction is given to first order in $S$ by

$$\psi_{AB} = c_A\psi_A + c_B\psi_B$$

where

$$c_A = \frac{1}{\sqrt{2}}\left[1 + \frac{\delta - S}{\sqrt{1 + \delta^2}}\right]^{1/2}$$

$$c_B = \frac{1}{\sqrt{2}}\left[1 - \frac{\delta + S}{\sqrt{1 + \delta^2}}\right]^{1/2}$$

with $\delta = \Delta E/2|h|$. Write down the corresponding electronic charge density and show that it leads naturally to definitions for the degree of ionicity $\alpha_i$ and the degree of covalency $\alpha_c$ that satisfy

$$\alpha_i^2 + \alpha_c^2 = 1.$$

3.2 Obtain the $\sigma$ and $\pi$ molecular orbital eigenstates for the $O_2$ dimer with a bond length of $R = 2.3$ au and s and p atomic energy levels $E_s$ and $E_p$ of $-29.1$ and $-14.1$ eV, respectively, using Harrison's matrix elements, namely

$$(ss\sigma, pp\sigma, sp\sigma, pp\pi) = (-2.80, 6.48, 3.68, -1.62)13.6/R^2 \text{ eV}.$$

Show that the oxygen dimer is predicted to be *paramagnetic* since the highest occupied level is a doubly degenerate $\pi$ state in which the two electrons will have parallel spin by Hund's rule. (This agrees with experiment unlike valence bond theory's prediction of diamagnetism due to all electrons being paired spin-up–spin-down.)

4.1 The interaction between noble gas atoms can be represented by the Lennard–Jones-type potential

$$\phi(R) = \varepsilon\left[\left(\frac{R_h}{R}\right)^{6\lambda} - \left(\frac{R_h}{R}\right)^6\right]$$

where $\alpha_h = (\lambda - 1)/\lambda$ defines the degree of normalized hardness of the interatomic potential. Plot this potential for $\alpha_h = \tfrac{1}{3}, \tfrac{1}{2}$, and 1, respectively, where $\alpha_h = \frac{1}{2}$ corresponds to the usual Lennard–Jones potential and $\alpha_h = 1$ corresponds to a hard-core potential.

Show that the equilibrium bond length for a tetrahedron of four noble gas atoms is given by

$$R_0^t(\lambda) = \lambda^{1/[6(\lambda-1)]}R_h$$

Hence, using the structural energy difference theorem compare the energies of the four-atom linear chain, square, and rhombus with that of the tetrahedron for values of the degree of normalized hardness $\alpha_h = \frac{1}{3}, \frac{1}{2}$, and 1, respectively. Comment on how the relative stability of the different four-atom clusters is affected as the interatomic potential becomes softer. Check that your results are consistent with the direct evaluation of the binding energy curves.

4.2   Find by diagonalizing the TB Hamiltonian matrix the eigenvalues for four s-valent atoms arranged as a linear chain, square, rhombus or tetrahedron, assuming only nearest neighbour bonding. Check your answer against the eigenspectra shown in Fig. 4.4 in units of $h_c$, the bond integral corresponding to cluster $c$. Use the structural energy difference theorem to evaluate the appropriate value of $h_c$ for each cluster in terms of that for the dimer $h_0$, assuming the distance dependence of the repulsive pair potential varies as the *square* of the hopping integral. Plot the bond energy per atom as a function of the number of valence electrons for the four different clusters. Using your predicted eigenspectra evaluate the moments $\mu_2$, $\mu_3$, and $\mu_4$ for each cluster. Why does each cluster have the same value of $\mu_2$? Check your predicted values for $\mu_3$ and $\mu_4$ by evaluating them directly by counting all paths of length 3 and 4 in the clusters respectively. Hence, explain why we find the structural trend from tetrahedron to rhombus to linear chain to square as the energy levels are occupied with electrons.

4.3   Discuss what is meant by the bond order. Why is it a powerful concept?

Derive by yourself the variation of the bond order with electron count that is shown in Fig. 4.8 for the case of four s-valent atoms that are configured as a linear chain, square, rhombus, and tetrahedron, respectively.

4.4   Find the eigenspectra of the trimer $AH_2$ as a function of the bond angle $2\beta$, neglecting any direct hydrogen–hydrogen bonding and ignoring the valence s orbitals on atom A. Why are the eigenfunctions either even or odd with respect to the mirror plane of the trimer? Show that within this approximation you expect $BeH_2$ to be linear whereas $H_2O$ should be bent with a bond angle of 90°. Why do you think water has the larger bond angle of 104°? (The more theoretical student can demonstrate the role of sp hybridization in increasing the bond angle above

90° by including the 2s orbital explicitly and finding the new eigenspectra and energy variation with bond angle, assuming Harrison's values for the bond integrals which are given in Question 3.2 using $R = 1.8$ au.)

5.1 Using the spherical jellium model explain the expected special stability of sodium clusters containing the 'magic number' of atoms 2, 8, 18, 20, 34, 40, 58, . . . .

5.2 Explain the concept of a pseudopotential. Aluminium is fcc with a lattice constant of $a = 7.7$ au. It is well described by an Ashcroft empty core pseudopotential of core radius 1.1 au. Show that the lattice must be expanded by 14% for the $2\pi/a(200)$ Fourier component of the pseudopotential to vanish.

5.3 The total energy (in Rydbergs) per atom of a NFE metal with valence $Z$ may be written to first order in the pseudopotential as $U = ZU_{eg} + U_{es}$ where

$$U_{eg} = \frac{2.21}{r_s^2} - \frac{0.916}{r_s}$$

and

$$U_{es} = \frac{-3Z^2}{R_{WS}}\left[1 - \left(\frac{R_c}{R_{WS}}\right)^2\right] + 1.2\frac{Z^2}{R_{WS}}$$

$r_s$, $R_c$, and $R_{WS}$ are the electronic, Ashcroft and Wigner–Seitz radii, respectively.

Derive the contribution $U_{es}$ by making the Wigner–Seitz sphere approximation, in which inter-cell electrostatic interactions are neglected and the intra-cell potential energy is approximated by that of the Wigner–Seitz sphere.

Show that at the equilibrium volume the bulk modulus

$$B = V\,\mathrm{d}^2U/\mathrm{d}V^2$$

may be written

$$B = (0.200 + 0.815R_c^2/r_s)B_{ke}$$

where

$$B_{ke} = 0.586/r_s^5.$$

Evaluate $B/B_{ke}$ for Na and Cu by fitting $R_c$ to their Wigner–Seitz radii of 3.99 and 2.67 au respectively (assuming $Z = 1$ in both cases). Compare with the experimental values of 0.80 and 2.16 respectively and comment on the large discrepancy for Cu.

6.1 The relative structural stability of a NFE metal is determined by an oscillatory pair potential of the type

$$\phi(R) = A\cos(2k_3R + \alpha_3)\frac{e^{-\kappa_3 R}}{R}$$

where $k_3 = 0.96 \, k_F$ and $\kappa_3 = 0.29 \, k_F$ ($k_F$ is the Fermi wave vector). Assuming that only the *twelve* first nearest neighbours contribute in a close-packed lattice (fcc or ideal hcp) and the *fourteen* first and second nearest neighbours in a bcc lattice, find the expected domains of bcc or close-packed stability within the $(Z, \alpha_3)$ structure map where $\alpha_3$ ranges from $-\pi$ to $\pi$ and $Z$ runs from 1 to 4. Assume that bcc takes the same equilibrium volume as close-packed. Comment on any differences from the structure map shown in Fig. 6.12.

6.2 Discuss the origin of the Hume–Rothery electron phases within the framework of Jones' original rigid-band analysis. How does second-order perturbation theory help quantify Mott and Jones' earlier supposition on the importance of the free electron sphere touching a Brillouin zone boundary?

7.1 Consider a simple cubic lattice of p valent atoms which form nearest neighbour bonds only. Show that the bandstructure $E(k, 0, 0)$ is given within the tight binding approximation by

$$E(k, 0, 0) = \begin{cases} E_p + 2pp\sigma \cos ka + 4pp\pi \\ E_p + 2pp\pi \cos ka + 2(pp\sigma + pp\pi) \\ E_p + 2pp\pi \cos ka + 2(pp\sigma + pp\pi) \end{cases}$$

where $E_p$ is the atomic energy level. Show that $pp\sigma$ is positive whereas $pp\pi$ is negative and that we expect $|pp\pi| \ll pp\sigma$. Hence, taking $pp\pi = 0$, plot the bandstructure from the centre of the Brillouin zone to the zone boundary at $\pi/a(100)$. Noting that the crystal has cubic symmetry, comment on why the bands are triply degenerate at $(0, 0, 0)$ but split into a single and a doubly degenerate level for $(k \neq 0, 0, 0)$.

7.2 Consider an infinite linear chain of sp-valent atoms lying along the $z$-axis with lattice spacing $a$. The atomic s and $p_z$ orbitals form strong $\sigma$ bonds between nearest neighbours along the chain. The $p_x$ and $p_y$ orbitals form much weaker $\pi$ bonds and may be neglected. The Bloch functions may, therefore be written as linear combinations of the atomic s and $p_z$ orbitals, namely

$$\psi_k(r) = c_s \psi_k^s(r) + c_p \psi_k^p(r)$$

where

$$\psi_k^\alpha(r) = \frac{1}{\sqrt{N}} \sum_R e^{ik \cdot R} \phi_\alpha(r - R)$$

where $\alpha \equiv$ s, $p_z$ and $k = (0, 0, k)$, $R = (0, 0, na)$ ($n$ an integer). Show that the resultant tight binding secular determinant takes the form

$$\begin{vmatrix} E_s + T_{ss}(k) - E(k) & T_{sp}(k) \\ T_{ps}(k) & E_p + T_{pp}(k) - E(k) \end{vmatrix} = 0$$

where

$$T_{\alpha\alpha'} = \sum_{R \neq 0} e^{ik \cdot R} \int \phi_\alpha(r) V_{\text{atom}}(r) \phi_{\alpha'}(r - R) \, dr.$$

Derive the matrix elements $T_{\alpha\alpha'}$ in terms of the nearest neighbour bond integrals, $ss\sigma$, $pp\sigma$ and $sp\sigma$, namely

$$T_{ss} = 2ss\sigma \cos ka,$$

$$T_{pp} = 2pp\sigma \cos ka,$$

and

$$T_{sp} = T_{ps}^* = 2i \, sp\sigma \sin ka.$$

Assume that $|ss\sigma| = pp\sigma = h$ and that the atomic s and p energy levels are the same and may be taken as the zero of energy (i.e. $E_s = E_p = 0$). Show that the bandstructure is given by

$$E(k) = \pm 2[sp\sigma^2 + (h^2 - sp\sigma^2) \cos^2 ka]^{1/2}.$$

Plot the bandstructure from $k = 0$ to $k = \pi/a$ in the *absence* of any sp-hybridization (i.e. $sp\sigma = 0$) and deduce that there is a *single* continuum of states of width $4h$. Show that in the *presence* of sp-hybridization of strength $sp\sigma = h$ the bandstructure splits into *two* flat bands which are separated by a hybridization or energy gap of $2sp\sigma$.

7.3   (i) Discuss the origin of the different contributions to the binding energy of transition metals that are indicated in Fig. 7.11. Show how the parabolic variation of the cohesive energy with band filling is accounted for by Friedel's rectangular d-band approximation.

(ii) Discuss the origin of the observed *negative* heats of formation of transition metal alloys with average d-band fillings around half-full but *positive* heats of formation for nearly empty or full d-bands.

7.4   Discuss the origin of the hybridization gap in sp-valent semiconductors. Show that within a simple model the hybridization gap exists provided

$$\Delta E_{sp} < 2|h|$$

where $\Delta E_{sp} = E_p - E_s$ and $h$ is the hybrid bond integral. Using Fig. 2.13 and Harrison's values for the bond integrals in Question 3.2, estimate the value of $\Delta E_{sp}/2|h|$ for C, Si, Ge, and Sn with respect to the diamond structure with lattice constant 6.75, 10.26, 10.70 and 12.27 au respectively.

8.1   (i) Show that the second moment approximation is unable to differentiate between the stability of different structure types within an s-valent nearest neighbour model with $\Phi(R) \alpha [h(R)]^2$.

(ii) Show that the fourth moment contribution from hopping out and back along three p-valent atoms with bond angle $\theta$ is given by

$$\mu_4^{(3)} = pp\sigma^4 \cos^2 \theta$$

for the case where $pp\pi = 0$.

Calculate the normalized fourth moments $\mu_4/\mu_2^2$ for the dimer, helical linear chain with bond angles of 90°, graphitic sheet with bond angles of 120°, and diamond lattice with bond angles of 109° (assuming $pp\pi = 0$). Amongst these structures which do you expect to be the most stable and the least stable for a half-filled p shell?

(iii) Explain why s-valent hydrogen is dimeric but sp-valent silicon is diamond-like. In group IV, why are the graphitic form of carbon and the close-packed form of lead more stable than diamond?

# Index

Where page numbers are given in *italic type*, they denote references to illustrations.